Study Guide:
Your Driver's Manual for Marketing

MARKETING
AN INTRODUCTION

Study Guide:
Your Driver's Manual for Marketing

MARKETING
AN INTRODUCTION

Eighth Edition

Gary Armstrong
Philip Kotler

Brandi Guidry Hollier
University of Loussiana-Lafayette

PEARSON
Prentice
Hall

Upper Saddle River, New Jersey 07458

VP/Editorial Director: Jeff Shelstad
Senior Acquisitions Editor: Katie Stevens
Product Development Manager: Ashley Santora
Associate Director, Manufacturing: Vincent Scelta
Production Editor & Buyer: Carol O'Rourke
Printer/Binder: Von Hoffman, Elridge

Pearson Prentice Hall™ is a trademark of Pearson Education, Inc.

10 9 8 7 6 5 4 3 2 1
ISBN 0-13-168716-6

Contents

Chapter 1   Marketing: Managing Profitable Customer Relationships 1

Chapter 2   Company and Marketing Strategy: Partnering to Build
Customer Relationships 12

Chapter 3   The Marketing Environment .24

Chapter 4   Managing Marketing Information 37

Chapter 5   Consumer and Business Buyer Behavior 50

Chapter 6   Segmentation, Targeting, and Positioning: Building the Right
Relationships with the Right Customers 67

Chapter 7   Product, Services, and Branding Strategy 81

Chapter 8   New-Product Development and Product Life-Cycle Strategies 99

Chapter 9   Pricing: Understanding and Capturing Customer Value 110

Chapter 10   Marketing Channels and Supply Chain Management 129

Chapter 11   Retailing and Wholesaling .146

Chapter 12   Communicating Customer Value: Advertising, Sales Promotion,
and Public Relations 158

Chapter 13   Communicating Customer Value: Personal Selling and Direct
Marketing 178

Chapter 14   Marketing in the Digital Age 196

Chapter 15   The Global Marketplace 211

Chapter 16   Marketing Ethics and Social Responsibility 227

Suggested Answers 241

Chapter 1
Marketing: Managing Profitable Customer Relationships

Previewing the Concepts—Chapter Objectives

1. Define marketing and outline the steps in the marketing process.
2. Explain the importance of understanding customers and the marketplace, and identify the five core marketplace concepts.
3. Identify the key elements of a customer-driven marketing strategy and discuss the marketing management orientations that guide marketing strategy.
4. Discuss customer relationship management and strategies for creating value *for* customers and capturing value *from* customers in return.
5. Describe the major trends and forces that are changing the marketing landscape in this age of relationships.

JUST THE BASICS

Chapter Overview

Marketing is all about customer relationships—profitable customer relationships. Whether your company sells expensive systems to a few key customers or toothpaste to millions, understanding the customer is the heart of a successful business. Although everyone within a company must be obsessed with making sure the customer is happy, one of the many roles of marketing is growing current customers and acquiring new ones.

Marketing is defined as *a social and managerial process by which individuals and groups obtain what they need and want through creating and exchanging value with others*. Figure 1-1 of the text shows a model of the marketing process that includes understanding the marketplace and customer needs and wants; designing a customer-driven marketing strategy; constructing a marketing program that delivers superior value; building profitable relationships and creating customer delight; and capturing value from customers to create profits and customer quality. The chapter reviews these five steps, focusing on the relationship aspect of marketing.

Chapter Outline

1. **Introduction**
   a. NASCAR is a great marketing organization. It is the second highest-rated regular season sport on television.
   b. It has a single-minded focus: creating lasting customer relationships.

    c.   A big part of the NASCAR experience is the feeling that the sport itself, is personally accessible. Because of this, NASCAR has attracted more than 250 big-name sponsors.

    d.   Today's successful companies have one thing in common: they are strongly customer focused and heavily committed to marketing.

2.   **What Is Marketing?**
    a.   A simple definition of marketing is *managing profitable customer relationships*.
    b.   Marketing must both attract new customers and grow the current customers.
    c.   Every organization must perform marketing functions, not just for-profit companies. Non-profits also must also perform marketing.

Marketing Defined
    d.   Most people think of marketing as selling and/or advertising, "telling and selling."
    e.   Marketing must focus on satisfying customer needs.
    f.   The formal definition of marketing is *a social and managerial process by which individuals and groups obtain what they need and want through creating and exchanging value with others*.

The Marketing Process
    g.   Figure 1-1 shows the five-step marketing process.
- Understand the marketplace and customer needs and wants.
- Design a customer-driven marketing strategy.
- Construct a marketing program that delivers superior value.
- Build profitable relationships and create customer delight.
- Capture value from customers to create profits and customer quality.

    h.   The first four steps create value for customers and build relationships with them.
    i.   The last step captures value from the customer in return for the value delivered.

3.   **Understanding the Marketplace and Consumer Needs**
    a.   Companies must first understand what the customer needs and wants, as well as the marketplace in which they operate.
    b.   Five core marketplace concepts are reviewed in this section:
- Needs, wants, and demands;
- Marketing offers (products, services, and experiences);
- Value and satisfaction;
- Exchanges, transactions, and relationships; and
- Markets.

Needs, Wants, and Demands

c.   Human *needs* are felt deprivation.

d.   They include physical needs (food, clothing, safety); social needs (belonging, affection); and individual needs (knowledge, self-expression).

e.   These needs are not created by marketing; they are intrinsic to humans.

f.   Human needs take the form of *wants* when culture and personality are applied. They are shaped by society.

g.   Wants become *demands* when they are backed by buying power.

h.   Value and satisfaction are the motives for people to demand products.

i.   Marketing research helps companies understand customers' needs, wants, and demands.

Marketing Offers—Products, Services, and Experiences

j.   A *marketing offer* is a combination of products, services, information or experiences offered to satisfy a need or want.

k.   Marketing offers can also include such things as persons, places, organizations, information, and ideas.

l.   *Marketing myopia* is paying more attention to the individual products offered, rather than the need satisfied, or benefits produced.

m.   Companies should focus on *brand experiences,* rather than just the product attributes.

n.   Experiences have emerged as differentiating factors for many companies, including Disney, Harley Davidson, and NASCAR.

Value and Satisfaction

o.   Consumers make their choices based on their perception of the value offered by each company.

p.   Companies have to be able to set the right level of expectations—set them too low, and they may succeed in satisfying them, but they won't be able to attract many customers. If they set expectations too high, they could risk disappointing customers.

Exchange, Transactions, and Relationships

q.   A core concept in marketing is *exchange,* which is the act of obtaining a desired object from someone by offering them something in return.

r.   The marketer tries to bring about a response to some market offering; the response may be more than buying or trading products and services.

s.   Marketing consists of actions taken to build and maintain desirable *exchange relationships* with target audiences.

Markets

t.   A *market* is defined as the set of actual and potential buyers of a product.

u.   Marketers must manage markets to create the desired exchange relationships.

v.   Sellers search for buyers, identify their needs, design good marketing offers, set prices for them, promote them, and store and deliver them.

w. Marketing generally involves serving a market of final buyers in the face of competitors.

x. A company's success depends on how well their entire system meets the needs of the consumer; the system includes their suppliers, their products and services, and any marketing intermediaries they use.

4. Designing a Customer-Driven Marketing Strategy

a. *Marketing management* is defined as the art and science of choosing target markets and building profitable relationships with them.

Selecting Customers to Serve

b. The company must decide whom it will serve by dividing the market into segments of customers *(market segmentation)* and selecting which segments to serve *(target marketing)*.

c. Some marketers need to reduce demand for their product; this is defined as *demarketing*.

Choosing a Value Proposition

d. A company must decide how it will *differentiate* and *position* itself.

e. A company's *value proposition* is the set of benefits or values it promises to deliver; these value propositions differentiate one brand from another.

f. Companies must design value propositions that give them advantage in the marketplace.

Marketing Management Orientations

g. Companies must decide on their *philosophy* to guide their marketing strategy.

h. There are five alternative philosophies.

i. The *production concept* says that consumers will favor products that are available and affordable.

j. In the *product concept,* consumers favor products that are highest in quality, performance, and innovative features.

k. Companies that utilize the *selling concept* undertake large-scale selling and promotional effort to get consumers to buy.

l. The *marketing concept* says that the company needs to understand the needs and wants of the target markets and deliver satisfaction better than their competitors do.

m. The *societal marketing concept* is a relatively new concept that asks companies to not overlook consumer long-run welfare while meeting their short-run wants.

n. Figure 1-3 illustrates the differences between the selling concept and the market concept.

5. **Preparing a Marketing Plan and Program**
   a. The marketing strategy outlines which customers the company will serve.
   b. Guided by that strategy, marketing programs are developed to deliver value to the target customers.
   c. The marketing mix is the set of tools the company uses to implement the strategy.
   d. The marketing tools are classified into four categories, called the *four Ps of marketing*: product, price, place, and promotion.

6. **Building Customer Relationships**
   a. The previous discussion covered the first three steps in the marketing process—understanding the marketplace and customer needs; designing a customer-driven marketing strategy; and constructing marketing programs.
   b. The fourth and most important step is building profitable customer relationships.

Customer Relationship Management
   c. *Customer Relationship Management (CRM)* is the overall process of building and maintaining profitable customer relationships by delivering superior customer value and satisfaction.
   d. A customer evaluates the difference between all the benefits and all the costs of a marketing offer; this is the *customer perceived value*.
   e. The perceived values and costs may not be accurate or objective.
   f. *Customer satisfaction* depends on the products' perceived performance relative to a buyer's expectations.
   g. The key is to match customer expectations with company performance.
   h. The marketer must balance customer satisfaction level with profitability.
   i. There are levels of customer relationships.
   j. The *basic relationship* level is at one extreme and is exhibited in markets with many low-margin customers.
   k. *Full partnerships* are developed in markets that have few customers and high margins.
   l. Many companies today develop customer loyalty and retention programs.
   m. One way of doing that is by offering *financial benefits*, such as *frequency marketing programs*.
   n. Companies can also add *social benefits*, such as *club marketing programs*.
   o. Yet another approach is *structural ties* such as Mc Kesson's online supply management system that helps retail pharmacy customers.

The Changing Nature of Customer Relationships
   p. Today's companies are building lasting relationships that are direct. They are targeting fewer, but more profitable customers.
   q. In addition to delivering value to customers, companies are assessing the value they get from customers.

r.   Companies are also trying to use CRM to build profitable, long-term relationships with customers that will enable them to retain the customers they wish to serve.

s.   Direct marketing is growing; some companies sell only through direct channels, such as Dell and Expedia.

Partner Relationship Marketing

t.   Companies work with many partners and engage in *partner relationship management*.

u.   Inside the company, every functional area could interact with customers; no longer is the marketing department solely responsible for understanding customers. Many companies are forming cross-functional selling teams, which can consist of sales and marketing people, operations specialists, financial analysts, and more.

v.   Outside the firm, the company may deal with distributors, retailers, and others. The *supply chain* stretches from raw materials to components to final products that are carried to final buyers.

w.   Connections must also be made with the members of the supply chain.

7.   **Capturing Value from Customers**

a.   In the last step in the marketing process, the company captures value from the customer.

b.   Satisfied customers stay loyal and buy more, which means greater long-run returns for the company.

Creating Customer Loyalty and Retention

c.   Completely satisfied customers are the most loyal, and even a slight drop in satisfaction can make a big difference in loyalty.

d.   *Customer lifetime value* is an important concept that says that if you lose a customer, you don't just lose one sale. You potentially lose thousands or even hundreds of thousands that a customer could spend over their lifetime.

e.   As a marketer, you need to help customers form an emotional bond with your brand, rather than just making a rational choice.

Growing Share of Customer

f.   Customer Relationship Management helps grow *share of customer*—the share the company gets of that customer's total spending on the product or service type.

g.   Companies can grow share of customer by offering greater variety to their current customers.

h.   Cross-selling and up-selling are two other ways of increasing customer expenditures with the company.

Building Customer Equity

i.   *Customer equity* is the total combined lifetime values of all the company's customers.

j.   Customer equity forecasts the future, whereas sales and market share tell what happened in the past.

k.   Companies can classify their customers as to whether they are profitable or not, and then manage the relationships accordingly.

l.   Figure 1-5 shows one model of customer classification by profitability and projected loyalty.

m.   A key learning is that different types of customers require different management strategies to maintain and increase profitability.

8.   **The New Marketing Landscape**

a.   In this section, five trends are identified and discussed that show how rapidly the business world is changing.

The New Digital Age

b.   Combined technology and information explosions have changed the way we relate to one another across the globe.

c.   Technology has enabled companies to learn more about customers, to get their products out to a much larger, global audience, and to tailor their products to individual customer needs.

d.   There are also many new ways to reach customers, including CD-ROMs, interactive TV, and other new technologies that let companies focus in on individual customer needs.

e.   The Internet is expected to reach almost 1.4 billion users by 2007.

f.   Business-to-business e-commerce was projected to reach $4.3 trillion this year compared with only $107 billion in consumer purchases.

Rapid Globalization

g.   Companies large and small are globalizing, if only because they are facing global competition themselves.

h.   This has resulted in a much more complex marketing environment for all companies.

i.   Companies are also buying more of their supplies outside their home country

j.   Many companies form strategic alliances and joint ventures with foreign companies to build global networks.

The Call for More Ethics and Social Responsibility

k.   Companies are being asked today to look at what their impact is on their environment.

l.   Those companies that look to the future are accepting their responsibilities in the areas of social and environmental impact.

m.   Many companies, such as Ben & Jerry's and others, practice "caring capitalism" by being civic-minded.

The Growth of Not-for-Profit Marketing

n.   Colleges, hospitals, museums, and other not-for-profit companies and associations are using the same marketing strategies as for-profit companies.

o.   Government agencies are also utilizing marketing for both recruitment and *social marketing campaigns.*

9.   So, What Is Marketing? Pull It All Together

a.   Marketing is the process of building profitable customer relationships by creating value for customers and capturing value in return.

b.   The first four steps are focused on creating value for the customer, while the last step returns value from the customer to the company.

c.   After the marketing strategy is defined, the marketing program is developed, which consists of the four Ps.

d.   When building value for customers, companies must utilize marketing technology, go global in both selling and sourcing, and act in an ethical and socially responsible way.

e.   Figure 1.6 shows a model of the marketing process, and the remainder of the text goes into detail on all of the concepts covered in this first chapter.

Creative Marketing Exercises Designed to Reinforce the Concepts!!!   (Suggested answers to these exercises can be found at the end of the Study Guide.)

1.   List 5 items that you perceive as "needs." What differentiates them from "wants?" Explain.

2.   Think of a product you were "offered" recently. Outline the manner in which it was presented to you and discuss why you did or did not accept this offer.

3.   A friend offers to make a trade with you. He offers you the guitar of your dreams for something of yours. What would you offer in exchange and why?

4.   Select your favorite frozen food. Toward whom do you think this product is targeted? Explain.

5.   You have created a new software program that will streamline the payroll process for small businesses. Which of the 5 marketing concepts will you use to design and sell your product? Justify.

6.   Think about your favorite video store and the last experience you had while there. What makes this store your favorite?

7.   Find 5 examples of companies partnering to sell products.

8.   How does your bank customize its services to meet the needs of its customers? Show examples.

9. Buick has a reputation for building customer loyalty from a specific demographic. Who are these customers and why are they so invested in Buick?

10. Go to www.cocacola.com and read the company's international mission statement. Who are its target customers?

"Linking the Concepts" (#1) -- Suggestions/Hints

1. Marketing can be paraphrased as the different ways in which individuals and companies "sell" goods and services to satisfy needs and desires to others in exchange for something else of value. Better definitions of marketing should include the value of long-term customer relationships through the exchange process.

2. Marketing informs consumers about the availability of goods and services that consumers might not have known about. Marketing is the tool in which companies sell their goods and services and thus create competition. The value of this competition is that prices are kept lower due to the competition. Marketing is experienced numerous times throughout a normal day. We experience numerous examples of marketing on TV, radio, forms of transportation, and during the sporting events throughout the day.

3. A consumer is more likely to purchase athletic shoes from a company that offers the product with the greatest perceived value and the greatest potential for satisfaction. In addition, societal interests, product pricing, promotional activities, and product availability all impact buying decisions.

"Linking the Concepts" (#2) -- Suggestions/Hints

1. Marketing is utilized to help deliver customer satisfaction, which will, hopefully, lead to long-term customer relationships and customer loyalty.

2. Lexus believes that if it delights a customer and continues to delight that customer, the company will have that customer for life. This philosophy is part of Customer Relationship Management. This can be viewed in a larger scope called full partnership as well. Wal-Mart does this by establishing relationships, not just with its customers, but also its suppliers.

3. Some local restaurants are really good at establishing relationships with their customers by learning what their regular customers like and how they like to have their meals prepared. This, and learning their customers' names, helps develop long-term relationships.

Marketing Adventure Exercises (Suggested answers to these exercises can be found at the end of the Study Guide.)

(Visit www.prenhall.com/adventure for advertisements.)

1.   General   Drugmart, Hallmark

Define "marketing offer" and identify the marketing offer in the selected ads.

2.   Travel   Imperial Hotel

The text describes the two important questions a marketing manager must answer to design a winning marketing strategy. What are those two questions and how do they apply to the selected ad?

3.   Exhibit   Student choice

Which organization in the Exhibits section might use "demarketing"?

4.   Financial   Student choice

Define value proposition and find an ad in the Financial section that clearly shows its value proposition.

5.   Student choice

Explain the differences between the product concept, the selling concept, and the marketing concept. Which categories contain advertisements that represent each of these philosophies?

6.   Student choice

Today, many companies are developing customer loyalty and retention programs by offering frequency marketing programs that reward customers who buy frequently or in large amounts. Which category of ads would most likely use frequency marketing programs? How would you alter the ads to incorporate frequency marketing?

7.   Financial   Student choice

Using selective relationship management, marketers are now targeting fewer, more profitable customers. Describe what selective relationship management is and which ads in the Financial section might employ such a strategy.

SUM IT UP!!!!!!

Using only this page, sum up all of the concepts and terms discussed in Chapter 1 – "Marketing: Managing Profitable Customer Relationships". Here is your chance to make sure you know and understand the concepts!!!

Chapter 2
Company and Marketing Strategy:
Partnering to Build Customer Relationships

Previewing the Concepts—Chapter Objectives

1. Explain companywide strategic planning and its four steps.
2. Discuss how to design business portfolios and develop growth strategies.
3. Explain marketing's role in strategic planning and how marketing works with its partners to create and deliver customer value.
4. Describe the elements of a customer-driven marketing strategy and mix, and the forces that influence it.
5. List the marketing management functions, including the elements of a marketing plan, and discuss the importance of measuring return on marketing.

JUST THE BASICS

Chapter Overview

This chapter looks at steps two and three of the marketing process—designing customer-driven marketing strategies and constructing marketing programs. It begins with a discussion of companywide strategic planning, something many students probably don't know even exists. But, as the chapter points out, marketing plans and programs are not developed in a vacuum. They must be a part of and consistent with the broader, overall strategic plans.

Next we discuss how marketers, guided by the strategic plan, work closely with others inside and outside the firm to serve customers. We then examine marketing strategy and planning ----how marketers choose target markets, position their market offerings, develop a marketing mix, and manage their marketing programs. Finally, we look at the important step of measuring and managing return on marketing investment.

Chapter Outline

1. **Introduction**
   a. Nike has forever changed the rules of sports marketing strategy.
   b. It has built the Nike swoosh into one of the world's best known brand symbols.
   c. The Nike of today is far different from the start-up company of 40 years ago.
   d. As Nike has grown and matured---moving from maverick to mainstream---its marketing strategy has matured as well.

e.   To stay on top in the competitive sports apparel business, Nike will have to keep finding fresh ways to bring value to customers.

2.   **Companywide Strategic Planning: Defining Marketing's Role**

a.   Strategic planning is *the process of developing and maintaining a strategic fit between the organization's goals and capabilities and its changing marketing opportunities.*

b.   Strategic planning is the basis for all of the rest of planning for the firm.

c.   Companies typically develop several types of plans: annual, long-range, and strategic.

d.   Annual and long-range plans focus on keeping their current businesses running.

e.   Strategic plans involve adapting the firm to take advantage of opportunities in its constantly changing environment.

f.   The company begins by defining its overall purpose and mission, which is outlined in Figure 2-1.

g.   The mission is turned into objectives that should guide the entire company.

h.   The best portfolio of businesses and products is defined based on the above.

i.   Then each business develops its own plans in support of the corporate strategic plan.

Defining a Market-Oriented Mission

j.   Businesses may begin to drift as they grow larger. Management should ask the questions: What is our business? Who is the customer? Who do customers value? What should our business be?

k.   These questions should be asked—and answered—frequently.

l.   A mission statement is *a statement of the organization's purpose—what it wants to accomplish in the larger environment.*

m.   Mission statements should be market oriented, not focused on the product(s) the company produces.

n.   This means that the mission should be defined in terms of satisfying basic human needs.

o.   The mission should neither be too broad or too narrow; they should be *realistic* and *specific*. They should fit the *market environment*. They should be based on *distinctive competencies*. They should be *motivating*.

Setting Company Objectives and Goals

p.   The mission must then be developed into detailed objectives. Each manager should have objectives they are responsible for reaching.

q.   Marketing strategies and programs are then developed to assist in meeting those objectives. The marketing strategies developed must be defined in greater detail, resulting in specific marketing programs.

<u>Designing the Business Portfolio</u>

r.   A business portfolio is *the collection of businesses and products that make up the company.* The best business portfolio is one that matches the company's strengths and weaknesses.

s.   Business portfolios are planned in two steps:
1.   The company analyzes their current businesses to decide which ones should receive more, less, or no investment.
2.   Then the future portfolio is developed through strategies for growth and/or downsizing.

t.   In portfolio analysis, each of the products and businesses making up the company is evaluated. Strong businesses and products should be supported, while weak ones should be phased down or dropped.

u.   A *strategic business unit* (SBU) is a part of the company that can be planned independently from other company businesses.

v.   The purpose of strategic planning is to find ways the company can match its strengths to opportunities in the environment.

w.   The two most important portfolio analysis methods measure two dimensions—the attractiveness of the SBUs market or industry, and the strength of the SBUs position in that market or industry.
1.   The Boston Consulting Group approach is the most well-known method of analysis.
        a.   In this method, the company's businesses are defined according to a growth-share matrix (see Figure 2.2)
        b.   The four types of SBUs are defined as follows:
                1.   Stars: high-growth, high-share businesses or products. These are expected to become Cash Cows.
                2.   Cash Cows: low growth, high-share businesses or products. They produce a lot of cash that the company uses to support SBUs that need investment.
                3.   Question Marks: low-share businesses in high-growth markets. They generally require a lot of cash, so management has to decide which ones to support and which should be shut down.
                4.   Dogs: low-growth, low-share businesses and products.
        c.   Once the SBUs are defined and classified in the matrix, the company has to decide what role each will play. There are four strategies that can be pursued:
                1.   The company can invest to build share.
                2.   The company can invest enough just to hold share.
                3.   The company can harvest, taking the cash for other uses.
                4.   The company can divest, or shut down the SBU.

        d.   SBUs and products can, and do, change their positions over time. The company must add new products and units all the time.

    2.   Matrix approaches have some limitations.

        a.   They are difficult and time-consuming. Defining and measuring market share and growth is especially difficult.

        b.   They also focus on what is happening today, rather than what should happen in the future.

        c.   They also can cause companies to go into areas unrelated to their strengths, simply because the market environment looks enticing.

    3.   Many companies today are decentralizing strategic planning, allowing cross-functional teams to do it.

x.   In addition to analyzing current businesses, companies must focus on profitable growth through finding businesses and products the company should move into in the future.

y.   Marketing is the function that identifies, evaluates and selects market opportunities, and develops strategies for going after them

z.   A way of doing this is through the *product/market expansion grid*, shown in Figure 2-3.

    1.   Market penetration means making more sales to current customers, without changing your product(s).

    2.   Market development is identifying and developing new markets for current products.

    3.   Product development is when the company offers modified or new products to current markets.

    4.   Diversification means moving into both new markets and new products, perhaps by starting up or buying businesses.

aa.   Companies also need to be concerned with strategies for downsizing businesses.

    1.   Products can become unprofitable, or the firm could have moved into markets with which it is unfamiliar.

    2.   When brands or businesses are unprofitable or no longer fit the overall strategy, a business should prune, harvest, or divest them.

    3.   Managers should focus on promising growth opportunities, not use up energy trying to salvage fading ones.

3.   Planning Marketing: Partnering to Build Customer Relationships

a.   Once strategic plans are in place, more detailed plans need to be developed.

b.   Marketing is key to this additional planning activity:

    1.   It provides the *guiding philosophy,* which is the marketing concept.

    2.   It provides *inputs to strategic planners* in the areas of attractive market opportunities and analyzing the company's capability for taking advantage of those opportunities.

3.   Within the business units, marketing *designs strategies* for reaching each unit's objectives.

c.   Marketing must work with all internal departments and external partners to attract, keep, and grow customers.

d.   Both *customer relationship management* and *partner relationship management* are important.

Partnering with Other Company Departments

e.   Each company department is a link in the company's *value chain*. That means that each department adds value in designing, producing, marketing, delivering, and supporting the company's products.

f.   Success depends on how well each department goes about adding value, and how well all the individual departments coordinate their work.

g.   In reality, the myriad departments in a company can be in conflict.
  1.   Marketing takes the customer's point of view, and works to increase customer satisfaction.
  2.   But that could cause other departments to do a poorer job, as they see their jobs to be defined.
  3.   Yet marketing is charged with getting everyone to "think customer."
  4.   Therefore, each department needs to understand the others, so that everyone can work together to achieve the company's objectives.

Partnering with Others in the Marketing System

h.   The company should not solely focus on its own value chain; it needs to also partner with its suppliers, distributors, and customers.

i.   *Value-delivery networks* are companies working together to improve their total performance.

j.   Today, competition can largely take place between entire value-delivery networks, rather than individual competitors.

4. **Marketing Strategy and the Marketing Mix**

a.   Figure 2-4 shows the major activities in managing marketing strategy and the marketing mix.
  1.   Consumers are in the center. Profitable customer relationships are the goal.
  2.   Marketing strategy is next—this is the broad logic under which the company attempts to develop profitable relationships.
  3.   Guided by the strategy, the company develops its marketing mix—product, price, place, and promotion.

Customer Centered Marketing Strategy

b.   Marketing requires a deep understanding of customers.

c.   There are many different kinds of consumers, and they exhibit many different kinds of needs. Companies cannot profitably serve them all.

d. To better define who they can serve, companies must divide up the total market. The three steps to do that are *market segmentation, target marketing,* and *market positioning.*

e. *Market segmentation* is the process of dividing a market into distinct groups of buyers with different needs, characteristics, or behavior who might require separate products or marketing programs.
   1. A market segment is a group of consumers who respond in a similar way to a given set of marketing efforts.

f. *Target marketing* involves evaluating each market segment's attractiveness, and then selecting one or more segments to enter.
   1. A company should target those segments that will return the greatest profitability while generating the greatest customer value.
   2. If there are limited resources, companies might want to pursue only one or a few special market niches.
      a. Major competitors will oftentimes ignore or overlook small niches.
   3. Other companies could serve related segments—these segments could have different kinds of customers but the customers have the same basic wants.
   4. Large companies may go after all market segments with a broad range of products.

g. *Market positioning* is the development of *a clear, distinctive and desirable place relative to competing products in the minds of consumers.*
   1. To develop a market position, the company must identify competitive advantages it has; that is, where they offer greater value than do their competitors.
   2. Effective positioning requires the company to actually deliver what it says it is delivering, and it must communicate the fact that it is delivering it.

Developing the Marketing Mix
h. The marketing mix is *the set of controllable, tactical marketing tools that the firm blends to produce the response it wants in the target market.*

i. The marketing mix is typically described as the "four Ps": *product, price, place,* and *promotion.*
   1. A *product* is the mixture of goods and services the company offers.
   2. *Price* is what the consumer must pay for the product.
   3. *Place* is the way the company makes the product available to customers.
   4. *Promotion* is the set of activities that the company uses to communicate the value of their product to the marketplace and to persuade consumers to buy it.

j. Effective marketing programs combine all these elements in a way that allows the company to achieve its objectives by delivering value to consumers.

k.	However, the four Ps take an internal view, rather than looking externally, or viewing the company and its products from the buyer's perspective.

l.	From this perspective, the four Ps become the four Cs:

    1.	*Product* becomes *Customer solution.*

    2.	*Price* becomes *Customer cost.*

    3.	*Place* becomes *Convenience.*

    4.	*Promotion* becomes *Communication.*

## 5.	Managing the Marketing Effort

a.	Management of the marketing process is highly important.

b.	Figure 2-6 shows the four important functions: *analysis, planning, implementation*, and *control.*

Marketing Analysis

c.	Analysis should be performed to understand the markets and marketing environment the company faces; company strengths and weaknesses; and current and future marketing actions to understand which opportunities the company can pursue.

    1.	SWOT analysis is used to evaluate the company's strengths (S), (W) weaknesses, opportunities (O), and threats (T).

    2.	Strengths include capabilities, resources, and positive situational factors. Weaknesses include negative internal factors and negative situational factors. Opportunities are favorable external factors and threats are unfavorable external factors.

Marketing Planning

d.	A detailed marketing plan has to be developed for each business, product or brand.

e.	Table 2-2 shows the major sections of a marketing plan for a product or a brand.

Marketing Implementation

f.	Marketing implementation turns *plans* into *actions*. It involves the activities that make the plans work.

g.	Implementation requires the company to blend its people, organizational structure, decision and reward systems, and company culture in a way that supports its strategies.

Marketing Department Organization

h.	The marketing organization must be designed such that it can carry out the strategies and plans that are developed.

i.	In small companies, one person may perform all the marketing functions. In large companies, many specialists are found.

j.	The *functional* organization is the most common form. This organizational form has the different activities headed by a functional specialist, such as sales, advertising, marketing research, etc.

k.	A *geographic* organization might be utilized in a company that sells nationally or internationally.

l.	A *product management* organization can be found in companies with many different products or brands.

m.	A *market* or *customer management* organization is used in companies that sell one product to many different kinds of markets and customers.

n.	Very large companies might utilize a *combination* of all these forms.

Marketing Control

o.	Results of marketing strategies and plans need to be evaluated, and where necessary, corrective action should take place.

p.	The control process includes the following:

    1.	*Operating control* checks the ongoing performance of the marketing programs against the annual plan.

        a.	If necessary, corrective action is taken.

        b.	The purpose is to ensure that the company sales, profits, and other goals are being met.

    2.	*Strategic control* looks at whether the company's basic strategies are matched to its opportunities.

        a.	A major tool to use here is a *marketing audit,* which is a comprehensive examination of a company's environment, objectives, strategies, and activities.

            1.	It covers all major marketing areas of a company, not just the problem areas.

            2.	The audit is conducted by an objective and experienced outside party.

Measuring and Managing Return on Marketing

q.	*Return on marketing (marketing ROI)* is the net return from a marketing investment divided by the costs of the marketing investment. See Figure 2.8.

r.	It measures the profits generated by investments in marketing activities.

s.	Marketing returns are difficult to measure.

t.	A company can assess ROI in terms of standard marketing performance measures such as brand awareness, sales, or market share.

u.	ROI is integral to strategic business decisions at all levels of the business.

Creative Marketing Exercises Designed to Reinforce the Concepts!!! (Suggested answers to these exercises can be found at the end of the Study Guide.)

1. Create a strategic plan for the next 5 years of your life. Determine where you are now, where you want to be in the future, and how you plan on getting there.

2. Find a copy of a mission statement for a dot.com company. Do you think its statement truthfully reflects its business practices and attitude toward customers? Explain.

3. Visit www.unicef.com and read the organization's mission statement. What is the vision of top management for this organization?

4. Identify a company that hosts several strategic business units (SBUs). What is the logic for mixing those particular organizations together? Explain.

5. Make a list of 3 companies that meet the description of each of the types of units in the Boston Consulting Group matrix. You should have 12 companies total.

6. Go to www.marriott.com and discuss how this organization has successfully diversified the hotel room.

7. Visit www.fritolay.com and list all the products in the snack value chain.

8. Describe a local company that uses segmentation to sell its products. Is it successful?

9. You have been hired to develop the marketing plan for a new antique store in your community. How would you begin this process?

10. The company you work for wants you to figure out why customer sales have dropped in the past few months. Use the marketing control process to speculate why there has been a decline.

"Linking the Concepts" -- Suggestions/Hints

1. In order to determine why companywide strategic planning is in a marketing textbook, we must first look at the definition of strategic planning. The textbook states that strategic planning is "the process of developing and maintaining a strategic fit between the organization's goals and capabilities and its changing marketing opportunities. It involves defining a clear company mission, setting supporting objectives, designing a sound business portfolio, and coordinating functional strategies." This definition describes the fact that everything a company does is connected to changing marketing opportunities. These opportunities are based on the company's customers, which means that the company must know how to communicate, market products, and facilitate transactions with its customers. These plans for handling opportunities need to be short term and long term, meaning a strategic thought process. Strategic planning for a company has to include marketing, due to the fact that all product, service, and customer transactions that are forecasted can only be accomplished through marketing.

2. Generally speaking, Starbucks' mission is to "fill souls, not bellies." The company strives to "change the way people live their lives, what they do when they get up in the morning, how they reward themselves, and where they meet." Its strategy to accomplish this "mission" is to "maintain phenomenal growth in an increasingly over-caffeinated marketplace. The tools used to accomplish Starbucks' strategies are more store growth, adding in-store products and features that get customers to stop in more often, stay longer, and buy more, develop new retail channels, developing new products and store concepts, increase international growth, and other examples such as hotels serving the Starbucks' brand. Each of the tools described are all tools of marketing.

3. Other functional departments at Starbucks are in support roles, strategically. The accounting/finance, real estate, international, management, and other departments must establish long-term plans that will help the corporation accomplish its growth through effective costing, purchasing, asset management, and other tasks. If done properly, this will help Starbucks reach all of its long-term goals. The marketers at Starbuck's must work side by side with the other departments to make sure that all strategic plans can realistically be reached. Can they grow within the current budgetary, personnel, and management limits? They can make sure that they maintain growth without making the product too expensive to customers, which would in turn, take away from overall customer value.

Marketing Adventure Exercises (Suggested answers to these exercises can be found at the end of the Study Guide.)

(Visit www.prenhall.com/adventure for advertisements.)

1.   Student choice

What is a "mission statement"? Select an ad and go to that company's website to read its mission statement. Compare the mission statement to the ad you have chosen. What are the parallels? Is the mission statement product oriented or market oriented?

2.   Student choice

What is a business portfolio? Select a large well-known company and visit their web site to discover their business portfolio. Is the advertised product an integral part of their business portfolio?

3.   Food   Perdue3

What are the four parts of a product market expansion grid? Consider the selected ad and determine what action the company might take to satisfy each element of the product market expansion grid.

4.   Cosmetics   Edge

Customer-centered marketing is a process that involves three steps: market segmentation, target marketing, and market positioning. Define these and then refer to the selected ad to describe these three steps for the chosen product.

5.   Financial   Advantage

Your text states, "In this age of customer relationships, the four P's might be better described as the four C's". Describe the four C's as they relate to this ad.

6.   Apparel   Hush Puppies

The selected ad clearly identifies the target market for the product. Who is that target and in which part of a marketing plan would decisions regarding this target be made?

7.   Electronics   Student choice

Identifying threats and opportunities is an important component of a marketing plan. Discuss the opportunities and threats that exist for products in the selected category.

SUM IT UP!!!!!!

Using only this page, sum up all of the concepts and terms discussed in Chapter 2 – "Company and Marketing Strategy: Partnering to Build Customer Relationships". Here is your chance to make sure you know and understand the concepts!!!!

Chapter 3
The Marketing Environment

Previewing the Concepts—Chapter Objectives

1. Describe the environmental forces that affect the company's ability to serve its customers.
2. Explain how changes in the demographic and economic environments affect marketing decisions.
3. Identify the major trends in the firm's natural and technological environments.
4. Explain the key changes in the political and cultural environments.
5. Discuss how companies can react to the marketing environment.

JUST THE BASICS

Chapter Overview

Although the first two chapters of the text provide an overview of all of the important topics in marketing and sets the stage for the remainder of the topics covered, this third chapter starts going into detail on the first step of the marketing process—understanding the environment in which the company operates.

The chapter describes the major micro- and macroenvironments in which the company operates. The microenvironments dealt with will build on the customer and partner relationships developed in prior chapters; they include the other company departments, as well those companies in the supply chain, the value chain, and the customers themselves. Interested publics are also discussed.

The macroenvironment includes demographic changes, and the economic, natural, technological, political, and cultural environment. All of these forces need to be studied continuously to ensure that the company's business and product portfolios are still meeting the needs of its customer base.

Chapter Outline

1. **Introduction**
    a. McDonald's has been losing share to what the industry calls "fast-casual" restaurants because consumers today want more choices.
    b. Americans are seeking healthier eating options. As the market leader, McDonald's often bears the brunt of criticism.
    c. McDonald's has strived to realign itself with the changing marketing environment, and it appears to be paying off.

d.   Marketers need to be good at building relationships with customers, others in the company, and external partners.

e.   A company's marketing environment consists of the actors and forces outside marketing that affect marketing management's ability to build and maintain successful relationships with target customers.

f.   More than any other group in the company, marketers must be the trend trackers and opportunity seekers. They have disciplined methods—marketing intelligence and marketing research—for collecting information about the marketing environment. They also spend more time in the customer and competitor environments

g.   A company's *marketing environment* consists of the actors and forces outside marketing that affect marketing management's ability to build and maintain successful relationships with target customers.

h.   There are both opportunities and threats in the marketing environment.

i.   The *microenvironment* consists of the actors close to the company that affect its ability to service its customers.

j.   The *macroenvironment* consists of the larger societal forces that affect the microenvironment.

2.   **The Company's Microenvironment**

a.   Figure 3-1 shows all of the players affecting the company from a micro point of view.

b.   Relationships with all these actors must be developed so that marketing management can successfully create customer value and satisfaction.

The Company

c.   All the interrelated functional groups within the company form the internal environment.

d.   Marketing management must take these other groups into account:
   1.   Top management sets the mission, objectives, broad strategies, and policies.
   2.   Finance finds the money to carry out the marketing plans.
   3.   R&D designs safe and attractive products.
   4.   Purchasing gets the supplies and materials needed.
   5.   Operations' produces and distributes the product.
   6.   Accounting measures revenues and costs, and helps marketing understand how well it is achieving objectives.

e.   All these departments must work in concert and according to the "marketing concept" to "think consumer."

Suppliers

f.   Suppliers are an important link in the company's value delivery system.

g.   Marketing managers must pay attention to the availability of supplies, because shortages, delays, and strikes could damage customer satisfaction.

h.   Suppliers today are frequently treated as partners in creating and delivering value to customers.

Marketing Intermediaries
i.   Marketing intermediaries help companies promote, sell, and distribute goods to final buyers.
j.   They include *resellers, physical distribution firms, marketing services agencies,* and *financial intermediaries.*
   1.   Resellers are distribution channel firms that help the company find customers or make sales to them. They include wholesalers and retailers, who buy and resell merchandise.
   2.   Physical distribution firms assist the company in stocking and moving goods from their points of origin to their destinations.
   3.   Marketing services agencies perform some of the marketing functions such as market research, advertising, and media selection and placement.
   4.   Financial intermediaries include banks, credit companies, insurance companies and others that help finance transactions or insure against risks.
k.   Marketing intermediaries are also important links in the value delivery system.

Customers
l.   There are five types of customer markets that must be studied.
   1.   *Consumer markets* are made up of individuals and households that buy goods and services for personal consumption.
   2.   *Business markets* buy the goods and services for further processing or for use in their production process.
   3.   *Reseller markets* buy goods and services to resell at a profit.
   4.   *Government markets* consist of government agencies that buy goods and services to produce public services, or transfer the goods to others who need them.
   5.   *International markets* are made up of and of the above types of customers in other countries.

Competitors
m.   Marketers must know their competitors' strengths so that they can develop positioning strategies that differentiate their own products against the competitors'.
n.   No single competitive strategy will work for all companies.

Publics
o.   A *public* is any group that has an actual or potential interest in or impact on an organization. There are seven types of publics:
   1.   *Financial publics* influence the company's ability to obtain funds.
   2.   *Media publics* carry news, features, and editorial opinions.
   3.   *Government publics* may develop and enforce regulations on product safety, truth in advertising and other matters.

4. *Citizen-action publics* are consumer organizations, environmental groups, minority groups, etc. that may question a company's decisions.

5. *Local publics* include neighborhood residents and community organizations.

6. *General publics* may be concerned about a company's products and activities.

7. *Internal publics* include workers, managers, etc. who need to feel good about their company.

3. **The Company's Macroenvironment**

   a. Figure 3-2 shows the macroenvironmental forces that affect a company in the way of shaping opportunities and posing threats.

   Demographic Environment

   b. *Demography* is the study of human populations in terms of size, density, location, age, gender, race, occupation, and other statistics.

   c. This is interesting to marketers because it involves people; it is people that make up markets.

   d. The world population totals 6.4 billion, and will pass 8.1 billion people by the year 2030.

   e. A growing population means growing human needs to satisfy. Market opportunities could also be growing if purchasing power is growing as well.

   f. Marketers track changing age and family structures, geographic population shifts, educational characteristics, and population diversity.

   g. In the United States, the single most important demographic trend is the changing age structure of the population.

      1. The *baby boomers* were born between 1946 and 1964, and account for 28 percent of the population. There are 78 million baby boomers, who have become one of the most powerful forces shaping the U.S. marketing environment.

      2. *Generation X* is a "birth dearth" generation, numbering 49 million people born between 1965 and 1976. They tend to be cautious in their economic outlook because they grew up in a time of recession and corporate downsizing.

      3. *Generation Y*'s members were born between 1977 and 1994, and number about 72 million. This generation is still developing their buying preferences and behaviors.

   h. Marketers must decide whether to develop marketing plans and strategies based on generational differences.

   i. The traditional family is being redefined.

      1. Married couples with children now make up only about 34% of the U.S. households; married couples without children make up 28%, single parents comprise 16%; 32% are nonfamily households.

27

2.   The number of working women has increased greatly from 1950 when it was about 30% of the U.S. workforce to just over 60% today.

j.   There are also great geographic shifts in populations, both between and within countries.

    1.   In the United States, there has been a shift toward the Sunbelt states over the last two decades.

    2.   Marketers are interested in these kinds of shifts because people in different regions buy differently.

    3.   The shift in where people live has also shifted where they work.

k.   The U.S. population is becoming better educated.

    1.   In 2003, 84% of the population over age 25 had completed high school, and 27% completed college, up from 69% and 17% in 1980.

    2.   The rising number of educated people will increase demand for quality products, books, magazines, travel, personal computers, and Internet services.

l.   There are also more white-collar workers in the United States.

m.   Ethnic and racial make up varies among countries.

    1.   Japan is at one extreme with the United States at the other.

    2.   The U.S. population is 68% white, 14% Hispanic, 13% African American, and 4% Asian; the remaining 1% is made up of American Indian, Eskimo and Aleut peoples.

    3.   Many companies design products and promotions to appeal to the diverse ethnic and racial groups.

n.   The gay and lesbian markets are also being recognized as important to marketers.

o.   Another attractive market segment is that of the 54 million people with disabilities.

p.   As the population grows more diverse, marketers will continue to diversify their marketing programs to take advantage of that diversity.

Economic Environment

q.   The *economic environment* consists of factors that affect consumer purchasing power and spending patterns.

r.   Nations vary greatly in their levels and distribution of income.

    1.   *Subsistence economies* consume most of their own agricultural and industries output. They represent few marketing opportunities.

    2.   *Industrial economies* are at the other extreme, and represent rich markets for many kinds of goods.

s.   Incomes change all the time, and marketers need to track those changes.

    1.   The 1990s saw a consumption frenzy fueled by income growth, federal tax reductions, rapid increases in housing values, and a boom in borrowing.

    2.   The 1990s then saw a recession hit, and consumers started to spend more carefully.

3. In the early 2000s, consumers face the age of the "squeezed consumer." Consumers must repay debts during a time of increased household and family expenses. They spend more carefully.

t. Marketers also need to pay attention to *income distribution*
1. At the top of income distribution in the United States are the *upper-class* consumers, who are generally not affected by current economic events.
2. The *middle class* is comfortable, but is somewhat careful in their spending.
3. The members of the *working class* stick to the basics of food, clothing, and shelter.
4. The *underclass* members are those on welfare and many retirees who must count pennies to make even the most basic purchases.

u. Ernst Engel, over a century ago, noted that people shifted their spending as their income rose. This is now known as *Engel's laws* and is noted in Table 3-1.

v. Changes in major economic variables have a large impact on the marketplace.

w. With adequate warning, companies can take advantage of changes in this environment.

Natural environment

x. The *natural environment* involves the natural resources that are needed as inputs by marketers, or that are affected by marketing activities.

y. Environmental concerns have grown over the last three decades. There arc several trends that should be tracked:
1. *Shortages of raw materials*: both renewable (forests, food) and nonrenewable (oil, coal, minerals) resources pose serious problems.
2. *Increased pollution* is a problem worldwide.
3. *Increased government intervention* in management of natural resources varies by country.

z. Companies are developing *environmentally sustainable* strategies and practices in an effort to develop an economy that can be supported indefinitely.

Technological Environment

aa. The *technological environment* is a dramatic force in the marketplace today creating new markets and opportunities.

bb. The United States leads the world in research and development spending, $312 billion in 2005.

cc. Many companies are adding marketing people to R&D teams to obtain a stronger marketing orientation.

dd. Safety is an increasing concern as technology becomes more complex.

1. The U.S. Food and Drug Administration (FDA) have regulations to test new drugs.
2. The Consumer Product Safety Commission sets safety standards for products.
3. Marketing must be aware of and adhere to regulations that affect developing new products.

Political Environment

ee. The *political environment* consists of laws, government agencies and pressure groups that influence or limit various organizations and individuals in a given society.

ff. Regulation can encourage competition and ensure fair markets.

gg. Governments develop *public policy* to help guide markets.

hh. Legislation affecting business has been increasing.
1. The United States has laws covering competition, fair trade practices, environmental protection, product safety, truth in advertising, consumer privacy, packaging and labeling, pricing, and other issues. Table 3-2 lists many of the most important laws.
2. The European Commission is also establishing a framework of laws covering many of these same issues.

ii. Business legislation is enacted to protect companies from each other; to protect consumers from unfair business practices; and to protect the interests of society.

jj. Government agencies have discretion in how they enforce the laws that are passed.

kk. Marketers need to track laws at the local, state, national, and international levels.

ll. Enlightened companies ask their managers to be socially responsible over and above existing laws.

mm. E-commerce has created an entirely new set of legal and ethical issues. Online privacy issues are a great concern.

nn. Many companies are exercising social responsibility through *cause-related marketing*.

Cultural Environment

oo. The *cultural environment* is made up of institutions and other forces that affect a society's basic values, perceptions, preferences, and behaviors.

pp. People are affected by the worldview that defines their relationships with others that their society adheres to.

qq. Cultural values are persistent; these beliefs shape specific attitudes and behaviors.
1. *Core* beliefs and values are passed on from parents to children.
2. *Secondary* beliefs are more open to change.

rr. Cultural swings do take place. Marketers want to predict these shifts in order to react to both opportunities and threats.

ss.   The major cultural values of a society are expressed in the following views:
1.   People's views of themselves.
2.   People's views of others.
3.   People's views of organizations.
4.   People's views of society.
5.   People's views of nature.
6.   People's views of the universe.

4.   Responding to the Marketing Environment
a.   Many companies think the marketing environment is an uncontrollable element to which they have to adapt.
b.   Other companies take an *environmental management perspective* to affect the publics and forces in their environment.
c.   Marketing managers should take a *proactive* rather than *reactive* approach to the marketing environment.

Creative Marketing Exercises Designed to Reinforce the Concepts!!! (Suggested answers to these exercises can be found at the end of the Study Guide.)

1.   Identify the factors in your classroom environment that affect your learning performance.
2.   Make a list of products that target senior citizens. Justify your choices.
3.   Think about your favorite reality television show. Why do you think these shows have become so popular?
4.   Conduct a demographic survey of your class. Use gender, age, race, ethnic background, and marital status as your starting point.
5.   Make a list of the members in your family that would be considered the "media public." Justify your choices.
6.   List five characteristics of a Generation Xer. How does the entertainment industry appeal to this market? Give examples.
7.   Create a spreadsheet depicting your budget for this week. How does the amount you have left over affect your weekend plans?
8.   Go to www.restaurant.org and locate the data depicting trends in eating out. What inference can you draw from this information?
9.   Visit the self-checkout aisle at your local grocery store. How does this technology better serve the customer?
10.   How might the war in Iraq affect the sale of American goods in that region?

"Linking the Concepts" (#1) – Suggestions/Hints

1. As an example of demographic developments that have impacted a person's life, there could be a student that was born during the last part of the Baby Boomer era. This person has moved from high school, where all the purchases acquired were for pleasure and enjoyment (stereo, concert tickets, lettermen's jackets) to now where his/her kids are graduating from college. This person is now buying more things for the kids and getting them ready to "take on the world". He/She is buying things for themselves, too, but these items are of bigger ticket prices (sports car, second home, vacation). Money is managed more at this person's age. He/She is now thinking about how the funds and lifestyle will be managed as they are looking at retirement, possibly. Returns on monetary investments are a major concern. As a younger person it was purely the "pleasure principle".

2. As examples of companies that are good at responding to the shifting demographic environment, Oldsmobile, McDonalds and Wal-Mart can be highlighted. Oldsmobile, for a long time, was the car company of older, financially set individuals. Years ago they developed a more youthful automobile and their campaign was based on the "This is not you father's Oldsmobile!" slogan. This was successful for them. Wal-Mart does advertisements that include many different diverse individual groups, such as Hispanic and the physically challenged. McDonalds, in an effort to respond to the changing demand for healthier foods, now offers more salads and nutritional information with their products.

An example of a company that could be perceived as a company slow to shift with the demographic environment might be IBM. They seem to take a long to put products on the market that are current with demographic needs. Their image is still relatively "stiff". They have maintained their old business image and work environment, where as Microsoft is perceived "fun" in its image and work environment.

"Linking the Concepts" (#2) Suggestions/Hints

1. As the economy changes, it has a direct impact on the amount consumers can spend on goods and services. If the economy starts to exhibit symptoms of prosperity, consumers tend to have more funds available to buy goods/services, and buy better quality items. If the economy shows signs of weakness, then consumers tend to hold on to their funds. During slow economical times, consumers tend to become value conscious. Fast food companies like McDonald's and Burger King have made major business gains with their value meal menus. Another economic effect connected to demographic forces is the fact that there are classes of income distribution. Wealthier groups look for higher priced, better quality goods. Groups, like the working class, tend to purchase only those items they need.

In terms of cultural trends, the values, perceptions, and behaviors of consumers have made major changes in marketing. People today are very time conscious. This causes consumers to look for goods and services that help save time. Fast food, drive through businesses, and stores open all night are examples of this trend. Also, there is a cultural trend toward more "stay at home and relax time." This is seen by the increase of TVs, DVD players, huge sound systems, and kitchens with time-saving features. There is also a cultural trend toward companies giving back to the community. McDonald's is a prime example of this by the use of the Ronald McDonald Children's Charities.

There is a large target market of consumers that are concerned about the natural environment. So this target market tends to shop for goods and services that are environmentally safe and made of biodegradable materials. Companies are even using their concerns about the environment when they produce advertisements that are designed to enhance their image.

Technology has helped make products better, cheaper, and easier to produce. This has helped make consumers lives more enjoyably and is easier on the wallet. One great example of an industry that has taken technology and created opportunities is the cell phone industry. Consumers can now purchase phones that take pictures, connect to the Internet, and also still make calls.

2. It is believed that the environments are both controllable and yet uncontrollable for marketers. In the sense that the environment is controllable, markets can stimulate demand for goods that were not exhibited. This is seen by every fashion design created for clothes. The environment did not force those changes. Marketing created those style changes through movies, TV, and other appearances.

In the sense that the environment is uncontrollable, marketing has to rely on government fiscal policy to help boost the economy. This will then allow consumers to have more money to purchase more, and better, goods and services.

Marketing Adventure Exercises (Suggested answers to these exercises can be found at the end of the Study Guide.)

(Visit www.prenhall.com/adventure for advertisements.)

1.   Household   Frux

Review the selected ad. Discuss the various marketing intermediaries, and their roles, that help bring this product from producer to consumer.

2.   Apparel   Levis4
      General   Worldbiggest

Companies study the various customer markets to provide goods and services to meet their needs. Who would probably be the customers for the products in the selected ads?

3.   Auto   Audi, Volvo

Your text states that "to be successful, a company must provide greater customer value and satisfaction than its competitors do." Examine the two selected ads for competing products. Which product appears to provide a greater value to the customer?

4.   Advertising   Shoot
      Auto   Hyundai
      Auto   Nissan3
      Financial   Berliner
      Nonprofit   Buenos_aires

A company's marketing environment includes various publics. Define a public, and describe which type of public the selected ads are aimed toward.

5.   Student choice

Refer to the discussion in your text regarding the changing age structure of the population. Based on that information, discuss the characteristics of each group (Baby Boomer, Generation X, Generation Y) and then select advertisements aimed at these targets.

6.   Cosmetics   Student choice

Your text discusses the changing American family, with married couples with children now making up only about 34 percent of the nation's 105 million households. Select one ad from the Cosmetics category that represents this move toward nontraditional households.

7.   Student choice

Two important characteristics of the U.S. population that are discussed in the text include the fact that people are better educated and are more likely to be employed in white collar jobs. Select an ad from various categories that would be aimed at this group in the workforce.

8.   Student choice

Today, many companies link themselves to worthwhile causes as a way to exercise their social responsibility and build more positive images. Which ads are examples of cause-related marketing?

SUM IT UP!!!!!!

Using only this page, sum up all of the concepts and terms discussed in Chapter 3 – "The Marketing Environment". Here is your chance to make sure you know and understand the concepts!!!!

Chapter 4
Managing Marketing Information

Previewing the Concepts—Chapter Objectives

1. Explain the importance of information to the company and its understanding of the marketplace.
2. Define the marketing information system and discuss its parts.
3. Outline the steps in the marketing research process.
4. Explain how companies analyze and distribute marketing information.
5. Discuss the special issues some marketing researchers face, including public policy and ethics issues.

JUST THE BASICS

Chapter Overview

As the textbook points out, to be successful, marketing managers must on a daily basis deal with mountains of marketing information. They need to understand what they need to know, when they need to know it, and how to find the information they need. And they must do all this cost-effectively.

This chapter reviews marketing information systems that work to get the right information, in the right form, at the right time to marketing managers so that they can make effective decisions. The marketing research process is also considered and outlined, as well as the use of marketing intelligence and internal data. Finally, other marketing information considerations are discussed. These include how small business and non-profit organizations use market research, the special problems encountered in performing international marketing research, and public policy and ethical considerations that need to be considered in marketing research.

Chapter Outline

1. **Introduction**
   a. Over 50 years, Coach developed a strong following for classically styled, high-quality leather handbags and accessories.
   b. In the early years, Coach didn't need a lot of marketing research because handbags were largely functional and women bought only two per year.
   c. By the mid-1990s, sales at Coach started to slow. As women entered the workforce, they needed different types of bags.

d. Women wanted more stylish and colorful bags, and high-end designers such as Gucci and Channel were responding to these trends.

e. It was time for a change, but Coach needed marketing research to gain a better understanding of the new handbag buyer.

f. Based on extensive marketing research, Coach overhauled its strategy and helped engineer a shift in the way women shop for handbags.

g. Research revealed that wanted more fashion in their handbags, so Coach launched the "Signature" collection.

h. Research led Coach to discover a "usage void" and design a small bag with a strap called the "wristlet" at prices as low as $38.

i. Research also let Coach to learn that were increasingly interested in non-leather bags. The company designed the "Hamptons Weekend" line of fabric bags.

j. Coach watches its customers closely, looking for trends that might suggest voids to fill.

k. Coach has achieved double-digit sales and earnings growth every period since 2000. Coach's stock price has jumped 940 percent since going public.

l. The MIS interacts with information users to assess information needs. Next, it develops needed information from internal databases, marketing intelligence activities, and marketing research. Then it helps analyze information to put it in the right form, and finally it distributes the information and helps managers use it to make good decisions.

2. **Assessing Marketing Information Needs**

   a. A good marketing information system must balance what users would like to have against what they really need and what is feasible to offer.

   b. This process begins by asking information users what they want. The MIS monitors the marketing environment so that it can provide decision makers with information they should have to make key decisions.

   c. Sometimes the information wanted cannot be provided, either because it is simply not available or because the MIS has limitations.

   d. Companies must monitor what it costs to obtain, process, store, and deliver information, because these costs can be quite high. The company must decide what the benefits are of obtaining additional information.

3. **Developing Marketing Information**

   a. There are several sources of marketing information; these include internal data, marketing intelligence, and marketing research.

   Internal Data

   b. Internal databases are electronic collections of information from data sources within the company. Marketing managers can readily access this information to identify marketing opportunities and problems, and to plan programs and evaluate performance.

c. There are many sources of internal data.
 1. The accounting department keeps records of sales, costs, and cash flows.
 2. The operations department has information on production schedules, shipments, and inventories.
 3. The marketing department may have information about customer demographics, psychographics, and buying behavior.
 4. The customer service department keeps records of customer satisfaction or service problems.
d. Internal data is usually easy to get access to, but has limitations. It was collected for other uses, so it may be incomplete or not in the form needed.

Marketing Intelligence
e. Marketing intelligence is the systematic collection and analysis of publicly available information about competitors and developments in the market place.
f. The goal of marketing intelligence is to improve decision making, assess and track competitors' actions, and provide early warning of opportunities and threats.
g. You can gather intelligence by talking to your own company employees, benchmarking competitors' products, researching the Internet, walking around trade show floors, and going through rivals' trash.
h. Companies can also get marketing intelligence from their suppliers, resellers, and key customers.
i. Companies can buy competitive products, monitor their sales, keep an eye out for new patents, and examine other physical evidence.
j. Reading competitors' annual reports and other SEC filings can provide a lot of information, as do business publications, trade show exhibits, press releases, advertisements, and web pages.
k. Online databases, many of which are free, are also a good source of information. The SEC website has all public company filings available, and the Patent Office has all patents and patent applications available online.
l. Subscription-based databases include Dialog, Hoover's, DataStar, LEXIS-NEXIS, Dow Jones News Retrieval, ProQuest, and Dun & Bradstreet's Online Access.
m. Many companies are providing competitive intelligence training to their employees to prevent the release of information.
n. Ethical questions come into play in gathering marketing intelligence. There is much available publicly, so there are no reasons today to "snoop." No one needs to break the law or the accepted code of ethics.

Marketing Research
o. Marketing research is the systematic design, collection, analysis, and reporting of data relevant to a specific marketing situation facing an organization.

p.   Marketing research can be used to understand customer satisfaction and purchase behavior, assess market potential and market share, or to measure effectiveness of pricing, product, distribution, and promotional activities.

q.   The marketing research process has four steps: defining the problem and research objectives; developing the research plan; implementing the research plan; and interpreting and reporting the findings.

1.   *Defining the Problem and Research Objectives*

a.   This is often the hardest step in the research process. The manager may know something is wrong, without knowing the specific cause.

b.   The objectives of the research are defined next. The three types of research objectives include *exploratory research*, which gathers preliminary information that helps define the problem and suggests hypotheses; *descriptive research*, which, simply, describes things such as market potential for a product, or demographics of the customer base; and *causal research*, to test hypotheses about cause-and-effect relationships.

c.   The statement of the research problem and objectives guides the entire research process.

2.   *Developing the Research Plan*

a.   Researchers now must define the exact information needed, develop the plan for gathering it, and present the plan to management.

b.   The research plan outlines sources of existing data, and then spells out the specific research approaches, contact methods, sampling plans, and instruments that will be used. This plan should be in the form of a written proposal. It can call for gathering secondary data, primary data, or both.

3.   *Gathering Secondary Data*

a.   Gathering secondary data should start with the company's internal data. There are also data reports that can be purchased (see Table 4-1) and commercial online databases.

b.   Secondary data can be obtained quickly and at lower cost than primary data. Some data available from secondary sources would not be available to or would be too expensive to collect for a single company.

c.   There are problems with secondary data. The data needed may not exist, or it may not be usable. The researcher must make certain that it is relevant, accurate, current, and impartial.

4. *Primary Data Collection*

   a. Primary data must often be collected. The same concerns about relevancy, accurateness, currency, and impartiality exist.

   b. Table 4-2 summarizes research approaches, contact methods, sampling plans, and research instruments that are available.

   c. Research approaches include observational research, which involves gathering primary data by observing people, actions, and situations, as well as ethnographic research, which observes people in the natural environment. Observational research can use mechanical methods of observation, such as Neilsen's people meters and checkout scanners.

   d. Survey research is the most widely used method for primary data collection and is the best approach for descriptive research. Single-source data systems start with surveys of consumer panels and continue with electronically monitoring their purchases and exposure to various marketing activities.

   e. Survey research is very flexible but also presents problems. People may be unwilling or unable to answer questions for a variety of reasons; they may answer even if they don't know an answer to appear smart; or they may try to give pleasing answers.

   f. Experimental research is suited for gathering causal information. Experiments involve selecting matched groups of subjects, giving them different treatments, controlling unrelated factors, and checking for differences in responses.

   g. There are many methods of contacting respondents to gather information. Mail questionnaires are used to gather large amounts of data. They are not very flexible, they may take longer to complete, and the response rate is usually low. On the other hand, respondents may give more honest answers, and no interviewers are present to potentially bias answers.

   h. Telephone interviewing is a very good method of gathering information quickly and is more flexible than mail questionnaires. Response rates tend to be higher. But the cost per respondent is very high, and people may not want to discuss personal questions with an interviewer. There is also the potential for interviewer bias. Interviewers could also record responses differently.

i. Personal interviewing can be done individually or in groups. Individual interviewing can be very flexible, but can cost three to four times as much as telephone interviews. Group interviewing is also called focus group interviewing. This involves inviting six to ten people to talk with a trained moderator. It has become one of the major methods of research, but it is hard to generalize from the results. The potential for interviewer bias is also big.

j. Focus groups can be held via teleconferencing, or even online.

k. A sample is a segment of the population selected to represent the population as a whole. The sample should be representative of the entire population so that the researcher can make accurate estimates of the thoughts and behaviors of the larger population.

l. Designing the sample involves three decisions: Who is to be surveyed? (called the sampling unit); how many people should be surveyed? (the sample size); and how should the people in the sample be chosen? (sampling procedure). Table 4-4 describes the different kinds of samples, which are probability sample and nonprobability samples.

m. There are two main research instruments—the questionnaire and mechanical devices.

n. The questionnaire is the most common instrument used; it can be administered in person, by phone, or online.

o. Closed-end questions include all possible answers from which subjects have to choose their response. Open-end questions allow respondents to answer in their own words.

p. Researchers need to be careful of the wording and ordering of questions. Simple, direct, and unbiased wording should be used, and questions should be in a logical order. Table 4-5 shows the many errors that can crop up.

q. Mechanical instruments include supermarket scanners and people meters. Other mechanical instruments measure physical responses of subjects.

5. *Implementing the Research Plan*

a. Implementing the plan involves collecting, processing, and analyzing the information.

b. The data collection phase of marketing research is usually the most expensive and most subject to error of any of the phases.

c. Researchers need to process the data to isolate important information. Data needs to be checked for accuracy and completeness. Results are tabulated and statistical measures are developed.

6. *Interpreting the Research Findings*
   a. The findings need to be interpreted, conclusions drawn, and reports made to management. Researchers should present findings that are useful in making decisions rather than focusing on raw data or statistical techniques.
   b. Managers and researchers should work together to interpret research results.

4. **Analyzing Marketing Information**
   a. Information analysis might involve analytical methods that will help marketers make decisions.
   b. These models will help answer the questions of "what if" and "which is best."

Customer Relationship Management (CRM)
   c. Smart companies collect information at every customer touch point. These touch points include customer purchases, sales force contacts, service and support calls, website visits, satisfaction surveys, credit and payment interactions, market research studies—every time a customer and the company are in contact.
   d. Customer Relationship Management consists of sophisticated software and analytical techniques that integrate customer information from all sources, analyze it in depth, and apply the results to build stronger customer relationships.
   e. CRM analysts develop data warehouses and use data mining techniques. A data warehouse is a companywide electronic storehouse of customer information. The purpose of a data warehouse is to integrate information the company already has. Data mining techniques are used to sift through the data to dig out interesting relationships and findings about customers.
   f. Companies can use CRM to understand customers better, provide higher levels of customer service, and develop deeper customer relationships. They can also use it to note high-value customers, target them more effectively, cross-sell the company's products, and create offers tailored to specific customer requirements.
   g. CRM systems can be very expensive to implement—U.S. companies will spend from $10 billion to $20 billion on software alone, yet more than half the CRM efforts fail to meet objectives. Most commonly, the failure occurs because companies see this as just a software or technology issue.
   h. When it works, CRM benefits far outweigh the risks and costs.

5. **Distributing and Using Marketing Information**
   a. The marketing information system must make the information available to managers and others who make marketing decisions or deal with customers.
   b. Many companies use an intranet to facilitate information distribution. The intranet provides ready access to data, stored reports, and so forth.

<table>
<tr><td>c.</td><td>Companies are increasingly allowing key customers and value-network members to access account and product information, along with other information. The systems that do this are called extranets.</td></tr>
</table>

6. **Other Marketing Information Considerations**

a. This section looks at marketing research in small businesses and non-profits, international marketing research, and public policy and ethics issues in marketing research.

Marketing Research in Small Businesses and Non-profit Organizations

b. Small organizations have the same information needs as larger firms. Start-up businesses need information about their markets, their industries, competitors, potential customers, and reactions to new offers. Existing small businesses need to track customer needs and wants, reactions to new products, and changes in the competitive environment.

c. Many marketing research techniques can be used in a less-formalized manner and at little or no expense.

d. Small businesses can gather good information by observing what is around them. Retailers can watch vehicle and pedestrian traffic to find areas in which to locate; companies can watch for competitor ads in local media; companies can visit competitor locations.

e. Small companies can conduct surveys using convenience samples; for instance, companies can invite small groups to lunch to discuss topics of interest. Retail salespeople can talk to customers in stores.

f. Small companies and non-profits can also perform simple experiments by changing themes in mailings and watching results.

g. Most of the secondary information available to large companies is also available to small companies and non-profits. Many associations publish data, and the U.S. Small Business Administration publishes a large number of reports.

International Marketing Research

h. International researchers follow the same steps as domestic researchers, but they face more and different problems. International researchers have to deal with differing markets in different countries.

i. Secondary data is often difficult to obtain. Most research firms that do international research operate in only a few countries.

j. When collecting primary data, it may be difficult to develop good samples outside the U.S. Data, and lists such as telephone directories, census information, and other data may be lacking in other countries.

k. Reaching respondents in other parts of the world can also be difficult. In some countries, few people have telephones; in others, the mail system is unreliable. Poor roads and transportation systems can make people difficult to reach, and few people in developing countries can access the Internet.

l.   Cultural differences, such as language, can be problematic. Translating questionnaires is difficult. Questionnaires should be re-translated back to English before being administered to be sure idioms, phrases, and statements don't take on unintended meanings.

m.   Consumers also differ in their attitudes toward marketing research. Some countries' customs prohibit talking to strangers; in others, research questions can be considered too personal.

n.   There can also be high illiteracy rates that keep people from responding.

Public Policy and Ethics in Marketing Research

o.   Most research benefits both companies and consumers. However, the misuse of marketing data can harm or annoy consumers.

p.   *Intrusions on Consumer Privacy*

1.   Most consumers feel positively about market research, but others resent it or even mistrust it.

2.   Sometimes consumers are "taken in" by market research that turns out to be attempts to sell them something. Other consumers confuse market research with telemarketing and say "no" before the interviewer can get started.

3.   A recent poll showed that 70% of Americans say that companies have too much of consumers' personal information, and 76% said that businesses had invaded their privacy. These concerns have led to lower response rates.

4.   The research industry is attempting to educate consumers about the benefits of marketing research and has adopted broad standards outlining researchers' responsibilities to respondents and the general public.

5.   Some companies are appointing a "Chief Privacy Officer" to safeguard privacy of consumers.

q.   *Misuse of Research Findings*

1.   In some cases, research surveys appear to be designed to produce the wanted effect. Most of this seems to be unintended rather than blatant misrepresentation. Researchers' choice of wording can have an effect on survey outcomes and conclusions.

2.   In other cases, supposedly independent research turns out to have been paid for by companies with an interest in the outcome.

3.   Each company must be responsible for policing their conduct and reporting of marketing research.

<u>Creative Marketing Exercises Designed to Reinforce the Concepts!!! (Suggested answers to these exercises can be found at the end of the Study Guide.)</u>

1. How accurate is the U.S. Census? How could this process be more effective and accurate?
2. Make a list of all the restaurants in a 5-mile radius. How is this information useful?
3. Call a local hotel and ask them how they conduct customer surveys. Do they feel this information is useful to their business?
4. Find a copy of a mail questionnaire and outline its advantages and disadvantages.
5. Visit www.landsend.com and locate the customer touch points on this website.
6. Conduct an Internet search for sites with international marketing research information. Which sites did you find to be the most helpful?
7. Go to www.courttv.com and take the "13th Juror" survey. How "up to date" is this information?
8. Is "dumpster-diving" an ethical practice? Why or why not?
9. List 5 misleading statements advertisers make about their products.
10. Create a course satisfaction survey for your marketing class and administer. What was the outcome?

<u>"Linking the Concepts" (#1) -- Suggestions/Hints</u>

1. Coach can use several different types of research to learn more about its customer's preferences and buying behavior. For example, observational research can be used to gather primary data by observing relevant people, actions, and situations. For example, consumers can be observed as they browse for purses, pick up and examine various purses, and make buying decisions. Also, survey research (descriptive research) could also be used to gather primary data. Coach can learn about people's knowledge, attitudes, preferences, or buying behavior by asking them directly.

    Marketing researchers usually draw conclusions about large groups of consumers by studying a small sample of the total consumer population. Coach will need to define their sampling unit, determine the sample size, and decide on the sampling procedure. Then, Coach will have to use a research instrument, perhaps the questionnaire, to gather information. The marketing research plan will then need to be put into action (i.e. collecting, processing, and analyzing the information). The research findings must then be interpreted and conclusions must be reported to management.

2. Perhaps a student could use the marketing research process to analyze their career opportunities. The first step in the marketing research process involves defining the problem and setting the research objectives. The problem could be defined as "the lack of a job" and the research objectives could be descriptive in nature {to describe the market potential for a product (the student) or the demographics and attitudes of customers (the employers)}. Research objectives must be translated into specific

information needs (i.e. the research might call for specific information such as consumer hiring patterns: who is being hired, what skills and education do they possess, etc.) Both primary and secondary data can be gathered. The third step calls for implementing the marketing research plan by gathering, processing, and analyzing the information. The fourth step consists of interpreting and reporting the findings.

"Linking the Concepts" (#2) -- Suggestions/Hints

1. The overall goal of the marketing information system(MIS) is to use people, equipment, and procedures to gather, sort, analyze, evaluate, and distribute needed, timely, and accurate information to marketing decision makers. The MIS first assesses information needs. The MIS primarily serves the company's marketing and other managers, but it may also provide information to external partners. Then, the MIS develops information from internal databases, marketing intelligence activities, and marketing research. Internal databases provide information on the company's own operations and departments. Such data can be obtained quickly and cheaply but often needs to be adapted for marketing decisions. Marketing intelligence activities supply everyday information about developments in the external marketing environment. Market research consists of collecting information relevant to a specific marketing problem faced by the company. Lastly, the MIS distributes information gathered from these sources to the right managers in the right form and at the right time.

2. When applying the MIS framework to Coach, it appears that the company is doing very well in the market research area. Coach watches its customer closely, looking for trends tat might suggest new market voids to fill. This year along, Coach will spend $3 million on marketing research, interviewing 14,000 women about everything from lifestyles to purse styles to strap lengths.

   When applying the MIS framework to Coach, it appears that the company does not seem to develop much information from internal databases. The chapter-opening story does not mention anything about Coach's use of internal databases to identify marketing opportunities and problems, plan programs, and evaluate performance.

Marketing Adventure Exercises (Suggested answers to these exercises can be found at the end of the Study Guide.)

(Visit www.prenhall.com/adventure for advertisements.)

1. Student choice

Information in the internal database can come from many sources. What are those sources? Select one ad and relate what internal information would be beneficial.

2.   Services   Varig

What are the four steps in the marketing research process? Review the selected advertisement and discuss what you believe was the problem or research objectives that initiated this ad?

3.   Newspaper   Student choice

Explain the difference between primary data and secondary data. Find an ad in the newspaper category that cites either primary or secondary data.

4.   Travel   Student Choice

Describe ethnographic research. Which ad in the Travel category would have benefited from this type of research?

5.   Nonprofit   Nike, Nike disabled, Nike disabled2

Marketers had to learn much about the target market to create this ad. Why might they have used observational research?

6.   Financial   Student choice

What is a focus group? Select ads in the Financial category that you believe might have used a focus group to better understand the target customer.

7.   Internet   Student choice

Your text states that the "latest technology to hit marketing research is the Internet." Which ad in the Internet category deals with research data?

8.   Financial   Student choice

Discuss the benefits of "customer relationship management" (CRM) and select an ad from the Financial category that may use CRM.

SUM IT UP!!!!!!

Using only this page, sum up all of the concepts and terms discussed in Chapter 4 – "Managing Marketing Information". Here is your chance to make sure you know and understand the concepts!!!!

Chapter 5
Consumer and Business Buyer Behavior

Previewing the Concepts—Chapter Objectives

1. Understand the consumer market and the major factors that influence buyer behavior.
2. Identify and discuss the stages in the buyer decision process.
3. Describe the adoption and diffusion process for new products.
4. Define the business market and identify the major factors that influence business buyer behavior.
5. List and define the steps in the business buying decision process.

JUST THE BASICS

Chapter Overview

Buying behavior is at the core of marketing: The knowledge and understanding of why we buy and how we buy should be the bedrock of every marketing program. This chapter covers both consumer and business buying behavior.

Consumer behavior incorporates concepts from both sociology and psychology. To understand consumers and their buying processes, therefore, is to understand the myriad influences encountered day in and day out. Cultural, social, personal, and psychological factors affecting buying behavior are explained. These factors clarify the *why* of buying.

The chapter also details the *how* of buying, by covering the consumer buying process. This process includes the stages of need recognition, information search, evaluation of alternatives, purchase decision, and post-purchase behavior. The chapter also details the buyer decision process for new products as well as providing information on consumer behavior internationally.

Business markets are also defined and analyzed. The differences between the consumer buying process and the business buying process are highlighted, as is the nature of the buying unit in business markets. Also discussed is business buying on the Internet.

Chapter Outline

1. **Introduction**
    a. Buyers of Harley-Davidson motorcycles are intensely loyal and devoted to the brand. Because of this, Harley-Davidson is at the top of the heavyweight motorcycle market.

b. Harley-Davidson's marketing managers spend a lot of time studying their buyers—they want to know who their customers are, what they think, how they feel, and why they buy a Harley rather than another brand.

c. The research revealed seven core types: adventure-loving traditionalists, sensitive pragmatists, stylish status seekers, laid-back campers, classy capitalists, cool-headed loners, and cocky misfits. Yet all Harley owners appreciated their bikes for the same basic reasons—independence, freedom, and power.

d. The example of Harley-Davidson shows that there are many factors that affect consumer buying behavior.

2. **Consumer Markets and Consumer Buying Behavior**

a. *Consumer buying behavior* refers to the buying behavior of final consumers—individual and households who buy goods and services for their own consumption. All of these consumers make up the consumer market.

b. The American consumer market consists of more than 296 million people who consume trillions of dollars' worth of goods and services each year. The global market consists of almost 6.4 billion people.

Model of Consumer Behavior

c. Most large companies research consumer buying decisions in great detail to answer questions about what consumers buy, where they buy, how and how much they buy, as well as when and why they buy.

d. Learning what, where, when, and how and how much they buy is easy, but understanding the *why* of buying is very difficult, because those reasons are usually locked deep inside the consumer's mind.

e. Understanding how consumers respond to varying marketing messages starts with the stimulus-response model of buyer behavior found in Figure 5-1.

f. Marketing stimuli consist of the four Ps of product, place, price, and promotion.

g. Other stimuli include major forces and events in the buyer's environment: economic, technological, political, and cultural.

h. All these inputs enter the consumer's black box and are then turned into responses such as product choice, brand choice, dealer choice, purchase timing and purchase amount.

i. The buyer's characteristics influence how he or she perceives and reacts to the stimuli, and the buyer's decision process itself affects the buyer's behavior.

Characteristics Affecting Consumer Behavior

j. Figure 5-2 shows the factors influencing consumer purchases. For the most part, marketers cannot control these factors, but they must always keep them in mind.

k. Cultural factors exert a broad and deep influence. Roles are played by the buyer's culture, subculture, and social class.

1. Culture is the most basic cause of a person's wants and behaviors. This behavior is largely learned from families and other important institutions.

2. A child born in the United States generally learns the values of achievement and success; activity and involvement; efficiency and practicality; progress; material comfort; individualism; freedom; humanitarianism; youthfulness; and fitness and health.

3. Marketers try to spot cultural shifts so that they can discover new wants and desires, and then develop new products to meet the new wants and desires. An example is the shift toward health and fitness, which created a huge industry.

4. Subcultures are groups of people with shared value systems based on common life experiences and situations. They include nationalities, religions, racial groups, and geographic regions. Four important subculture groups in the United States include Hispanics, African Americans, Asians, and mature consumers.

    a. The U.S. Hispanic market, which includes Americans of Cuban, Mexican, Central American, South American, and Puerto Rican descent, numbers 42 million consumers who bought more than $686 billion worth of goods and services. This group is expected to double in size in the next 20 years. They tend to buy more branded, higher-quality products, and are extremely brand loyal.

    b. African Americans number 39 million with a buying power of $723 billion. This population is growing in affluence and sophistication. They can be more price-conscious than other groups, but they are motivated by quality and selection. Black consumers are also the most fashion-conscious of the ethnic groups.

    c. Asian Americans are the fastest growing and most affluent segment in the Untied States. They number 12.5 million and have a disposable income of $363 billion annually. The largest group consists of the Chinese Americans, followed by Filipinos, Japanese Americans, Asian Indians, and Korean Americans. Asian Americans are the most tech-savvy segment, and they shop frequently and are the most brand-conscious of all the ethnic groups. But they are also the least brand loyal.

    d. Mature consumers number 75 million, and this group

    e. will more than double in size in the next 25 years. Mature consumers are better off financially than are the younger consumer groups.

5. Social classes are society's relatively permanent and ordered divisions whose members share similar values, interests, and

behaviors. There are seven American social classes, and they are outlined in Figure 5-3.

6. Social class is not determined by a single factor; rather, it is measured by a combination of occupation, income, education, wealth, and other variables.

7. Marketers are interested in social class because people within a given class can exhibit similar buying behavior. Social classes have distinct product and brand preferences in areas such as clothing, home furnishings, leisure activity, and cars.

8. *Social Factors*
   a. A consumer's behavior is affected by social factors such as small groups, family, and social roles and status.
      1. Behavior can be influenced by small groups. These groups include membership groups, to which a person belongs; reference groups, which are indirect points of comparison or reference; and aspirational groups, to which an individual would like to belong. These reference groups expose a person to new behaviors and lifestyles, influence the person's attitudes and self-concept, and create pressures to conform that may affect a person's product and brand choices.
      2. Opinion leaders are those people within a reference group who exert influence on others. Buzz marketing is used by enlisting or even creating opinion leaders to spread the word about specific brands.
      3. Family members can influence buyer behavior. It is the most important consumer buying organization in society, and marketers study the roles of husbands, wives, and even children on the purchases of different products and services.
      4. Women make almost 85% of all purchases, totaling $6 trillion each year. Children also can have strong influence on family buying decisions.
      5. A person's position in each group can be defined in terms of role and status. A role consists of the activities people are expected to perform according to the persons around them. Each role carries a status reflecting the general esteem given to it by society. People often choose products that show their status in society.

9. *Personal Factors*
   a. Personal characteristics that affect what a consumer buys include age and life-cycle stage, occupation, economic situation, lifestyle, and personality and self-concept.

1. People change the goods and services they buy over their lifetimes. The family life cycle consists of the stages through which families might pass as they mature over time. Traditional family life cycle stages include young singles and married couples with children. Other alternative stages include unmarried couples, singles marrying later in life, childless couples, same-sex couples, single parents, those recently divorced, and extended parents (those with young adult children returning home).

2. A person's occupation can also affect what he or she buys, as will his or her economic situation.

3. A person's lifestyle is his or her pattern of living as expressed in his or her psychographics. It involves measuring consumers' major AIO dimensions—activities, interests, and opinions. Lifestyle profiles a person's whole pattern of acting and interacting with the world.

4. There are lifestyle classifications, the most popular of which was developed by SRI Consulting. It is called the Values and Lifestyles (VALS) typology. It classifies people according to how they spend their time and their money. It divides people up into eight groups based on two major dimensions: primary motivation and resources. Primary motivations include ideals, achievement, and self-expression. Resources are classified as either high or low, and include income, education, health, self-confidence, energy, and other factors.

5. Forrester research has also developed a "technographics" scheme, which segments consumers according to their motivation, desire, and ability to invest in technology.

6. Personality refers to the unique psychological characteristics that lead to relatively consistent and lasting response to one's own environment. Personality generally covers such traits as self-confidence, dominance, sociability, autonomy, defensiveness, adaptability, and aggressiveness. Personality influences buying behavior.

7. It is posited that brands also have personalities, and consumers will buy brands whose personality matches their own. A brand personality is the specific mix of human traits that may be attributed to a particular brand.

10. *Psychological Factors*

    a.   There are four psychological factors that could influence buyer behavior. They are motivation, perception, learning, and beliefs and attitudes.

        1.   A motive or drive is a need that is sufficiently pressing to direct the person to seek satisfaction. A person has many needs; some are biological, such as hunger and thirst; some are psychological, such as the need for recognition and esteem.

        2.   There are several theories of human motivation, but two of the most popular were developed by Sigmund Freud and Abraham Maslow.

            i.   Freud assumed that people are unconscious about the real psychological forces shaping their behavior, and that a person does not really understand his or her motivation.

            ii.   Maslow believed that people are driven by particular needs at particular times, and that needs are arranged in a hierarchy (see Figure 5-4). These needs include physiological needs, safety needs, social needs, esteem needs, and self-actualization needs. A person focuses on his or her most important needs, and as each level of needs ceases to be a motivator, he or she moves up the hierarchy.

        3.   Perception is the process by which people select, organize, and interpret information to form a meaningful picture of the world. How a person acts, or buys, is influenced by his or her perception of the situation. There are three perceptual processes through which people can form different perceptions of the same stimulus:

            i.   Selective attention is the tendency for people to screen out most of the information to which they are exposed.

            ii.   Selective distortion describes the tendency of people to interpret information in a way that will support what they already believe.

            iii.   Selective retention shows that people tend to retain only information that supports their attitudes and beliefs.

            iv.   Subliminal advertising refers to marketing messages received without consumers knowing it. Studies find no link between subliminal messages and consumer behavior.

4.   Learning describes changes in an individual's behavior arising from experience. It occurs through the interplay of drives, stimuli, cues, responses, and reinforcement.

5.   A belief is a descriptive thought that a person has about something. An attitude describes a person's relatively consistent evaluations, feelings, and tendencies toward an object or an idea. Attitudes put people into a frame of mind of liking or disliking things, and of moving toward or away from them. Attitudes are difficult to change.

The Buyer Decision Process

l.   There are five stages in the buyer decision process (see Figure 5-5): need recognition, information search, evaluation of alternatives, purchase decision, and post-purchase behavior. In more routine purchases, consumers can skip or even reverse some of these stages.

1.   Need recognition is the start of the decision process. The buyer recognizes that he or she has a problem or need.

a.   The need can come from external stimuli, such as advertising, or internal stimuli, such as hunger or thirst.

2.   Information search may or may not take place. If it does, consumers can get their information from many sources.

a.   Personal sources include family, friends, neighbors, and so forth.

b.   Commercial sources are advertising, salespeople, packaging, and so forth.

c.   Public sources include the media and consumer-rating organizations.

d.   Experiential sources include the consumer handling or examining the product itself.

e.   The relative importance of each of these sources varies by consumer. The most effective sources tend to be personal; commercial sources tend to inform the consumer, but personal sources can legitimize the products.

3.   Evaluation of alternatives depends on the individual consumer and the buying situation. In some case, consumers can evaluate product alternatives very carefully, using careful calculations and logical thinking. At other times, very little consideration of alternatives is done—impulse buying and relying on intuition rule.

4.   The purchase decision entails forming the purchase intention. Typically, consumers will now buy what they have decided on. However, two factors can come between the purchase intention and purchase decision.

        a.   First, attitudes of others can intervene. If someone close to the consumer casts doubt on the decision made, the purchase might not take place.

        b.   Second, there could be unexpected situational factors. A consumer could see a drop in income, or a competitor could drop their prices.

    5.   Post-purchase behavior is also considered part of the buying process. The difference between the consumer's expectations and the perceived performance of the good purchased determines how satisfied the consumer is. If the product falls short of expectations, the consumer is disappointed; if it meets expectations, the consumer is satisfied; if it exceeds expectations, the consumer is said to be delighted.

        a.   Cognitive dissonance generally results from every major purchase. This is the discomfort caused by post-purchase conflict. Every purchase involves compromise, through forgoing benefits of other products.

m.   It always costs more to gain new customers than to retain existing customers, and the best way to retain those you already have is to satisfy them. Bad word of mouth travels far more quickly than good.

The Buyer Decision Process for New Products

n.   A new product is a good, service, or idea that is perceived by some potential customers as new.

o.   The adoption process is the mental process through which an individual passes from first learning about an innovation to final adoption, and *adoption* itself is defined as the decision by a consumer to become a regular user of the product.

p.   There are five stages in adopting a new product.

    1.   Awareness is when the consumer becomes aware of a new product, but still lacks information about it.

    2.   Interest occurs as the consumer seeks information.

    3.   Evaluation is the process through which the consumer considers whether trying the new product makes sense.

    4.   Trial of the new product on a small scale improves his or her estimate of its value.

    5.   Adoption occurs when the consumer decides to make full and regular use of the new product.

q.   There are adopter categories into which consumers fall. See Figure 5-6.

    1.   Innovators tend to be adventuresome. They will try new ideas at some risk.

    2.   Early adopters are typically guided by respect. They are opinion leaders and adopt new ideas carefully, although early.

    3.   The early majority are rarely leaders, but they do adopt new ideas before the average person.

        4.   The late majority are skeptical; they wait for a majority of people to adopt something new before they do.

        5.   Laggards are tradition-bound. They can be suspicious of changes and adopt "new" ideas only when they've become somewhat of a tradition themselves.

r.   The characteristics of a new product affect its rate of adoption. Five characteristics are especially important in an innovation's rate of adoption.

        1.   Does it have a relative advantage over existing products?

        2.   Compatibility with the values and experiences of potential customers is important.

        3.   The degree of complexity is also considered. That is, how difficult is it to understand or use the product?

        4.   Divisibility speaks to how easily the innovation may be tried on a limited basis.

        5.   Communicability is the degree to which the results of using the innovation can be observed or described to others.

        6.   Other characteristics that influence adoption of new products are the initial and ongoing costs, the risk and uncertainty involved, and the level of social approval.

Consumer Behavior Across International Borders

s.   Understanding the needs of consumers across borders is a daunting task. Consumers in other countries may have many of the same things, but their values, attitudes, and behaviors often vary greatly.

t.   Marketers must decide the degree to which they will adapt their products and marketing programs to meet unique cultures and needs in various markets. Standardizing offerings simplifies operations and allows companies to take advantage of cost economies. But adapting marketing efforts within each country results in products and programs that better satisfy the needs of local consumers.

3.   Business Markets and Business Buyer Behavior

a.   Most large companies sell to other organizations. Even large consumer-products companies must first sell their products to other businesses before consumers can buy them.

b.   Business buyer behavior refers to the buying behavior of the organizations that buy goods and services for use in the production of other products and services that are sold, rented, or supplied to others. It includes the behavior of retailing and wholesaling firms that acquire goods for the purpose of reselling or renting them to others at a profit.

c.   In the business buying process, business buyers determine which products and services their organizations need to purchase, and then find, evaluate, and choose among alternative suppliers and brands.

Business Markets

d.   The business market is huge. Business markets involve more dollars and items than do consumer markets. There are many sets of business purchases for each set of consumer purchases.

e.   Business markets differ in many ways from consumer markets. The main differences are in market structure and demand; the nature of the buying unit; and the types of decisions and the decision process involved.

1.   *Market structure and demand*

       a.   The business marketer generally deals with far fewer but far larger buyers than the consumer marketer does. Even in large business markets, a few buyers often account for most of the purchasing.

       b.   Business markets are more geographically concentrated. Eight states account for more than half the nation's business buyers: California, New York, Ohio, Illinois, Michigan, Texas, Pennsylvania, and New Jersey.

       c.   Business demand is derived demand—it ultimately derives from the demand for consumer goods. Because of this, business marketers may promote their products directly to final consumers to increase business demand.

2.   *Nature of the buying unit*

       a.   A business purchase usually involves more decision participants and a more professional purchasing effort. Business buying is often done by professional purchasing agents.

       b.   The more complex the purchase, the more likely that several people will participate in the process. Buying committees made up of technical experts and top management are common in buying major goods.

3.   *Types of decisions and the decision process*

       a.   Purchases often involve large sums of money, complex technical and economic considerations, and interactions among many people at many levels of the buyer's organization. Because of the complexity, business buying decisions can take longer than consumer decisions.

       b.   The business buying process is generally more formalized than the consumer process. Detailed product specifications, written purchase orders, careful supplier searches, and formal approval are usually required.

       c.   In business buying situations, buyer and seller are often much more dependent on one another. They may work closely together, partnering to jointly create solutions to the customer's problems.

Business Buyer Behavior

f.   The business buyer behavior model, shown in Figure 5-7, shows how marketing and other stimuli affect the buying organization and produce certain buyer responses. As in consumer buying, the marketing stimuli for business consists of the four Ps: product, price, place, and promotion. Other stimuli include environmental forces such as economic, technological, political, cultural, and competitive forces. The stimuli are turned into buyer responses, such as product or service choice, supplier choice, order quantities, and delivery, service, and payment terms.

g.   There are three major types of buying situations.

    1.   In a straight rebuy, the buyer reorders something without any modifications. It is generally handled on a routine basis by the purchasing department.

    2.   In a modified rebuy, the buyer wants to modify product specifications, prices, terms, or suppliers. The modified rebuy usually involves more decision participants than does the straight rebuy.

    3.   A new task situation is encountered when a company is buying a product or service for the first time.

    4.   The buyer makes the fewest decisions in the straight rebuy and the most in the new task situation.

h.   Systems selling is often a key business marketing strategy because many business buyers prefer to buy a packaged solution to a problem from a single seller. In this situation, a buyer may ask sellers to supply the components and assemble the package or system.

i.   The decision-making unit of a buying organization is called the buying center. It is comprised of all the individuals and units that participate in the process. It is not a fixed and formally identified unit within the buying organization. Different people assume different roles for different purchases. For some purchases, only one person will participate; for more complex purchases, the buying center could include 20 or 30 people.

j.   Business buyers are subject to many influences when they make their buying decisions. They respond to both economic and personal factors.

k.   The various influences on buyers are shown in Figure 5-8. They include environmental, organizational, interpersonal, and individual influences.

    1.   Environmental factors can include the current and expected economic environment, as well as shortages of key materials. Technological, political, and competitive developments can also affect business buyers. Culture and customs can also influence buyer reactions to the marketer's behavior and strategies.

    2.   Organizational factors are important because each buying organization has its own objectives, policies, procedures, structure, and systems.

    3.   Interpersonal factors also influence the business buying process. These can be very difficult to ascertain.

4. Individual factors are involved as well. Each participant in the business buying process brings in personal motives, perceptions, and preferences. These are, in turn, influenced by personal characteristics such as age, income, education, professional identification, personality, and attitudes toward risk.

l. There are eight stages to the business buying process, which are shown in Figure 5-9. Buyers who are facing a new-task situation will usually go through all stages of the buying process. Those going through modified or straight rebuys may skip some of the stages.

1. Problem recognition: The buying process begins when someone in the company recognizes a problem or need that can be met by acquiring a specific product or service.

2. A general need description is generated to describe the characteristics and quantity needed of an item. For complex items, buyers may need to work with others, such as engineers, users, and consultants, to define the item.

3. The product specification includes the technical product specifications. Value analysis is an approach to cost reduction in which components are studied carefully to determine if they can be redesigned, standardized, or made by less costly methods of production.

4. A supplier search is conducted to find the best vendors. The newer the buying task, and the more complex and costly the item, the greater the amount of time the buyer will spend searching for suppliers.

5. The proposal solicitation is the stage in which the buyer invites qualified suppliers to submit proposals. When the item is complex or expensive, the buyer will usually require detailed written proposals or formal presentations from each potential supplier.

6. Supplier selection occurs after the buying center reviews the proposals. During the selection process, the buying center may draw up a list of desired supplier attributes and their relative importance. Buyers may also attempt to negotiate with preferred suppliers for better prices and terms before making final selections.

7. An order-routine specification is now prepared that includes the final order with the chosen supplier or suppliers. It lists items such as technical specifications, quantity needed, expected time of delivery, return policies, and warranties. Buyers may use blanket contracts for routine items; this creates a long-term relationship in which the supplier promises to resupply the buyer as needed at agreed prices for a set period of time.

8. Finally, buyers conduct a performance review. In this stage, the buyer may contact users and ask them to rate their satisfaction.

This review may lead the buyer to continue, modify, or drop the arrangement with the seller.

E-Procurement: Buying Electronically and on the Internet

m.   Online purchase, sometimes called e-procurement, is growing rapidly. Companies can set up their own company buying sites, create extranet links with key suppliers, or create well-designed easy-to-use Web sites.

n.   E-procurement gives buyers access to new suppliers, lowers purchasing costs, and hastens order processing and delivery.

o.   Business marketers can connect with customers online to share marketing information, sell products and services, provide customer support services, and maintain ongoing customer relationships.

p.   Most products bought online are MRO materials—maintenance, repair, and operating items. MRO materials account for up to 80% of all business orders, and the transaction costs for order processing are very high.

q.   Online procurement in the business-to-business environment shaves transaction costs and results in more efficient purchasing for both buyers and suppliers. E-procurement reduces the time between order and delivery and frees purchasing people to focus on more strategic issues.

There are some problems with e-procurement, however. There is the potential for security disasters; the secure environment that businesses need to carry out confidential interactions is still lacking.

Creative Marketing Exercises Designed to Reinforce the Concepts!!!   (Suggested answers to these exercises can be found at the end of the Study Guide.)

1.   What impact do you think you have on the U.S. economy? Global economy? Explain.

2.   Look through your favorite magazine and make a list of those items that are considered "trendy."

3.   Visit www.atkinsdiet.com and give a brief background on this controversial diet plan. At what point did this plan change from being a "fad" to a "trend?"

4.   Identify a radio station in your listening area that caters to a subculture. What specific strategies do they use?

5.   Outline 5 ways your family influences your buying. Be specific.

6.   Conduct a brief interview with a new mother. Ask about her diaper preference and what persuaded her to choose this particular brand.

7.   Make a list of your favorite television shows and write a brief synopsis on the lifestyle they depict.

8.   Go to www.checksunlimited.com and discuss how this site is an example of how advertisers appeal to a person's personality to sell a product.

9.   Find 5 examples of advertisements that use perception to sell a product.

10.   Using Figure 5.5, outline the process you used in purchasing the last piece of clothing you acquired.

"Linking the Concepts" – Suggestions/Hints

1. An example of a recent purchase to view the consumer buying decision process might be a new car. In the need recognition step, it could have been brought on by a recent auto accident that required a new car. As the consumer moved to the information search step, the consumer could think about those sources he/she already knows, internally, about where to purchase the vehicle and what products are available. He/she may not know of many sources to make the purchase, so he/she will need to seek out information concerning sources and products available from advertisements, for example. In the evaluation of alternatives stage, the consumer would start to examine what he/she would like in his/her purchase. Internal factors, such as motivation, and external factors, such as social influences, would now be evaluated to assist in determining which product to purchase. In the purchase decision stage, the consumer would make the purchase. Does he/she finance, pay all at once, trade in, or what? In the post-purchase behavior step, the consumer would look for those things that help reassure him/her that he/she made the correct decision.

2. If a person looks at NASCAR, it can be seen that this organization understands its customers and it uses that knowledge to build quality relationships with those customers. For example, race fans are known to be very loyal to their drivers, the cars, and the sponsors that are on the vehicles and around the tracks. If, for example, a person is a Tony Stewart fan, then that fan will tend to make sure all home improvement products will be purchased at Home Depot. NASCAR will then take that loyalty of that driver and apply it to other product promotions, and then those fans will tend to be brand loyal to those new products because they sponsor their driver.

3. A company that does business- to-business marketing will tend to use the same concepts of marketing as if they were marketing to final consumers. Both companies will be concerned with delivering quality products and services and meeting the needs of their customers. The only difference is that one company is selling to another business and the other company will be concerned about marketing to the final consumers. Granted, a business buying from a business may have more specific and detailed needs that are related to cost, delivery, and not so much as to pretty colors, like consumers may be. Still, both companies are meeting customer needs. Business-to-business companies use the same buying process as consumers; it just has a few more steps that are related to cost, quality, and relationships with suppliers.

Marketing Adventure Exercises (Suggested answers to these exercises can be found at the end of the Study Guide.)

(Visit www.prenhall.com/adventure for advertisements.)

1.   Student choice

Consumer purchases are strongly influenced by cultural factors. What cultural factors have influenced you? Select an ad that signifies these influences.

2.   Services, Nonprofit   Student choice

From time to time, cultural shifts will affect consumer behavior. Which products in the Services and Nonprofit categories are results of a recent cultural shift?

3.   Auto   Student choice

Subcultures and social class have a major influence on consumer buying. Choose two ads from the Auto category that are based on either subculture or social class.

4.   Apparel, Nonprofit, Travel   Student choice

Define lifestyle and discuss its various components. Choose ads in the selected categories that match the components you have listed.

5.   Auto   Audi, Chrysler, Fiat, Jaguar, Mercedes, Nissan, Toyota

Review the five levels in Maslow's Hierarchy of needs. Consider the motivation behind the selected ads and match them to the appropriate level in Maslow's hierarchy of needs. Discuss why these levels were easy to find while ads using physiological and self-actualization needs as the motivation would be more difficult.

6.   Electronics   Sony6

What is cognitive dissonance? Would purchasing products in this category likely cause cognitive dissonance? Does the selected ad seem to recognize this occurrence?

7.   Internet   Student choice

Define business buying behavior and discuss the ways in which it differs from the consumer market. Select an ad in the Internet category that is aimed specifically at the business market.

8.   Services   Corts, Hsd, Sedex, Speedstart, Varig

What are the three major types of buying situations? Match the selected ads to each of these situations and explain your rationale.

SUM IT UP!!!!!!

Using only this page, sum up all of the concepts and terms discussed in Chapter 5 – "Consumer and Business Buyer Behavior". Here is your chance to make sure you know and understand the concepts!!!!

Chapter 6
Segmentation, Targeting, and Positioning:
Building the Right Relationships with the Right Customers

Previewing the Concepts—Chapter Objectives

1.   Define the three steps of target marketing: market segmentation, market targeting, and market positioning.
2.   List and discuss the major bases for segmenting consumer and business markets.
3.   Explain how companies identify attractive market segments and choose a target marketing strategy.
4.   Discuss how companies position their products for maximum competitive advantage in the marketplace.

JUST THE BASICS

Chapter Overview

Market segmentation and target marketing are detailed in this chapter. An overview of consumer segmentation variables, including geographic, demographic, psychographic and behavioral characteristics, is explained, as is the use of multiple segmentation bases. Business market segmentation and international market segmentation are also described. The importance and reasons for segmentation are portrayed through several examples of companies who do it well, most notably Proctor & Gamble.

Segmentation outlines the company's opportunities, but target marketing is where the marketing manager makes his or her money. Turning the segmentation opportunities into real markets is the focus of this part of the chapter. Methods of evaluating the market segments are discussed, as are the various levels of targeting: undifferentiated or mass marketing; differentiated marketing; concentrated marketing; and micromarketing.

Choosing the target marketing strategy is described as dependent on many variables, such as company resources, how variable the product is, and the stage of the product life cycle. But while a company is targeting important segments, it must take care not to cause any controversy or concern. For instance, companies that have targeted their premium cereals primarily to children have been called to task for their practices, as have cigarette companies that seem to have targeted the youth market in their chosen advertising vehicles.

Finally, a company has to figure out the best way to position for competitive advantage. Positioning involves implanting the brand's unique benefits and differentiation in customers' minds. But how does the company do that effectively? This section of the chapter goes through the process of developing a positioning concept and statement, and describes the various ways to decide how to correctly position your product in the marketplace.

Chapter Outline

1. **Introduction**
   a. Proctor & Gamble is one of the world's premier consumer goods companies. They provide a good example of how smart marketers use segmentation, targeting, and positioning.
   b. Proctor & Gamble sells eight brands of laundry detergent in the United States, six brands of hand soap, five brands of shampoo, four brands of dishwashing detergent, three brands of tissues and towels and deodorant, and two brands each of fabric softener, cosmetics, skin care potions, and disposable diapers.
   c. The reason they do this is because different people want different mixes of benefits from the products they buy. There are groups—or segments—or laundry detergent buyers, for example, and each segment seeks a special combination of benefits.
   d. By segmenting the market and having several different brands in each category, P&G has an attractive offering for consumers in all important preference groups.
   e. Companies recognize that they cannot appeal to all buyers in the marketplace, or at least not to all buyers in the same way. So they must design strategies to build the right relationships with the right customers.
   f. Most companies are being more choosy about the customers with whom they wish to build relationships. They have moved away from mass marketing and toward market segmentation and targeting—identifying market segments, selecting one or more of them, and developing products and marketing programs tailored to each.
   g. The three steps in target marketing are shown in Figure 6-1. They are market segmentation, target marketing, and market positioning.

2. **Market Segmentation**
   a. Markets consist of buyers. These buyers may differ in their wants, resources, locations, buying attitudes, and buying practices. Through market segmentation, companies divide large, heterogeneous markets into smaller segments that can be reached more effectively with products and services that match their unique needs.
   Segmenting Consumer Markets
   b. There is no single way to segment a market. Marketers must try different segmentation variables, alone and in combination, to find the best way to view the market structure.
      1. Geographic segmentation divides the market into different geographical units, such as nations, regions, states, counties, cities, or even neighborhoods. A company may operate in one or a few

geographic areas, or it may operate in all areas but pay attention to geographical differences in wants and needs.

2. Demographic segmentation divides the market into groups based on variables such as age, gender, family size, family life cycle, income, occupation, education, religion, race, generation, and nationality. Consumer wants, needs, and usage rates often vary with demographic variables. Demographic variables are also easier to measure than other variables.

3. Some companies use age and life-cycle segmentation because consumer needs and wants change with age. But marketers must be careful to guard against stereotypes when using this form of segmentation.

4. Gender segmentation has long been used in clothing, cosmetics, toiletries, and magazines.

5. Income segmentation has been used by marketers for selling automobiles, boats, clothing, cosmetics, financial services, and travel. Many companies target affluent consumers with luxury goods. But other companies target lower-income consumers. Others still develop different products and sell them in different outlets based on income segmentation.

6. Psychographic segmentation divides buyers into different groups based on social class, lifestyle, or personality characteristics. People in the same demographic group can have very different psychographic makeups, and marketers often segment by common lifestyles.

7. Behavioral segmentation divides buyers based on their knowledge, attitudes, uses, or responses to a product.

   i. Occasion segmentation groups buyers according to occasions when they get the idea to buy, actually make the purchase, or use the purchased item.

   ii. Benefit segmentation requires finding the major benefits people look for in the product class, the kinds of people who look for each benefit, and the major brands that deliver each benefit.

   iii. User status groups buyers according to whether they are nonusers, ex-users, potential users, first-time users, or regular users of the product.

   iv. Markets can also be segmented according to usage rate— light, medium, and heavy product users.

   v. Loyalty status looks at the level of loyalty to brands, stores, and companies.

c. Marketers rarely limit their segmentation to only one or a few variables. They are increasingly using multiple segmentation bases in an effort to identify smaller, better-defined target groups.

1. Geodemographic segmentation helps marketers link U.S. Census data with lifestyle patterns to better segment their markets down to zip codes, neighborhoods, and even city blocks.

Segmenting Business Markets

d. Business marketers use many of the same variables to segment their markets. Business buyers can be segmented geographically, demographically (industry, company size), or by benefits sought, user status, usage rate, and loyalty status.

e. Other characteristics are also used, however, including operating characteristics, purchasing approaches, situational factors, and personal characteristics.

f. Within a given target industry and customer size, the company can segment by purchase approaches and criteria. Many marketers believe that buying behavior and benefits provide the best basis for segmenting business markets, just as in consumer markets.

Segmenting International Markets

g. Different countries can vary greatly in their economic, cultural, and political makeup. International firms need to group their world markets into segments with distinct buying needs and behaviors.

h. Companies can segment international markets using one or a combination of several variables. They can segment by geographic location. This assumes that countries close to one another will have many common traits and behaviors.

i. World markets can also be grouped on the basis of economic factors, such as population income levels or by their overall level of economic development.

j. Countries can also be segmented by political and legal factors, such as the type and stability of government, receptivity to foreign firms, monetary regulations, and the amount of bureaucracy.

k. Cultural factors can also be used, grouping markets according to common languages, religions, values and attitudes, customs, and behavioral patterns.

l. Many companies use an approach called intermarket segmentation. Using this approach, they form segments of consumers who have similar needs and buying behavior even though they are located in different countries.

Requirements for Effective Segmentation

m. Not all segmentations are effective. To be useful, segments must meet five criteria.
   1. It must be measurable: The size, purchasing power, and profiles of the segments can be measured.
   2. It must be accessible: The market segments can be effectively reached and served.

3.   It must bc substantial: The segments are large or profitable enough to serve. A segment should be the largest possible homogenous group worth pursuing with a tailored marketing strategy.

4.   It must be differentiable: The segments are conceptually distinguishable and respond differently to different marketing mix elements and programs.

5.   It must be actionable: Effective programs can be designed for attracting and serving the segments.

3. **Target Marketing**

a.   Segmentation reveals only the firm's opportunities. The firm now has to evaluate the various segments and decide how many and which segments it can best serve.

Evaluating Market Segments

b.   A firm must look at three factors to evaluate market segments: segment size and growth; segment structural attractiveness; and company objectives and resources

c.   The company must first collect and analyze data on current segment sales, growth rates, and expected profitability for various segments. It will be interested in segments that have the right size and growth characteristics. But "right size and growth" is a relative matter.

d.   There are several structural characteristics that affect long-run segment attractiveness.

1.   The segment is less attractive if there are several strong, aggressive competitors.

2.   The existence of many actual or potential substitute products may limit prices and the profits that can be earned.

3.   The relative power of buyers also affects segment attractiveness.

4.   A segment may be less attractive if it contains powerful suppliers who can control prices or reduce the quality or quantity of ordered goods and services.

e.   The company must take into account its own objectives and resources in relation to the segment. If a segment does not mesh with the company's long-run objectives, it can be dismissed. The company must take into consideration whether it has the skills and resources needed to succeed in the market. The company should enter only segments in which it can offer superior value and gain advantage over competitors.

Selecting Target Market Segments

f.   A target market consists of a set of buyers who share common needs or characteristics that the company decides to serve.

g.   Target marketing can be carried out at several different levels. Figure 6-2 shows that companies can target very broadly, through undifferentiated marketing; very narrowly, in micromarketing; or somewhere in between, which is differentiated or concentrated marketing.

71

1.   In undifferentiated marketing, also called mass marketing, a firm might decide to ignore market segment differences and target the whole market with one offer. This strategy focuses on what is common in the needs of consumers, rather than on what is different.

2.   Using differentiated, or segmented, marketing, a firm decides to target several market segments and designs separate offers for each. Companies hope for higher sales and a stronger position within each market segment. This could yield more total sales than undifferentiated marketing across all segments. Differentiated marketing can also increase costs, however. So companies must weigh increased sales against increased costs when deciding on a differentiated marketing strategy.

3.   Concentrated or niche marketing is especially appealing when a company has limited resources. Instead of going after a small share of a large market, the firm goes after a large share of one or a few segments or niches. Niches are smaller than segments and may attract only one or a few competitors. A company can market more effectively by fine-tuning its products, prices, and programs to the needs of carefully defined segments.

4.   Micromarketing is the practice of tailoring products and marketing programs to suit the tastes of specific individuals and locations. It includes local marketing and individual marketing.

    i.   Local marketing entails tailoring brands and promotions to the needs and wants of local customer groups—cities, neighborhoods, and even specific stores. The drawbacks include that it can drive up manufacturing and marketing costs, and create logistics problems as companies try to meet the varied requirements of the different markets. A brand's image may also be diluted and the message could vary too much. But local market does help a company market more effectively to different segments.

    ii.   Micromarketing becomes individual marketing in the extreme—tailoring products and marketing programs to the needs and preferences of individual customers. Mass customization is the process through which firms interact one-on-one with masses of customers to design products and services made specifically to individual needs. The move towards individual marketing mirrors the trend in consumer self-marketing, in which consumers take more responsibility for determining what to buy.

    iii.   Which strategy to employ depends on company resources.

        a.   When the firm's resources are limited, concentrated marketing makes sense.

        b.   Undifferentiated marketing makes sense when product variability is low, such as in steel.

       c.   The product's life-cycle stage must also be considered. When a product is new, undifferentiated marketing might be best. In the mature stage, differentiated marketing could work better.

       d.   Market variability needs to be considered, as should competitors' strategies.

Socially Responsible Target Marketing

       h.   Target marketing sometimes generates controversy and concern. Issues usually involve the targeting of vulnerable or disadvantaged consumers with controversial or potentially harmful products. Problems arise when marketing adult products to kids, whether intentionally or unintentionally.

       i.   The growth of the Internet and other carefully targeted direct media has raised concerns about potential targeting abuses.

       j.   The issue is not so much who is targeted, but how and for what. Controversies arise when marketers attempt to profit when they unfairly target vulnerable segments or target them with questionable products or tactics.

       k.   Socially responsible marketing calls for segmentation and targeting that serve not just the interests of the company, but also the interests of those targeted.

4.   Positioning for Competitive Advantage

       a.   A product's position is the way the product is defined by the consumers on important attributes; it is the place the product occupies in consumers' minds relative to competing products. It involves implanting the brand's unique benefits and differentiation in customers' minds.

       b.   To simplify the buying process, consumers organize products, services, and companies into categories and "position" them in their minds. A product's position is a complex set of perceptions, impressions, and feelings that consumers have for the product compared with competing products.

       c.   Consumers will position products with or without the help of marketers. So marketers must plan positions that will give their products the greatest advantage in selected target markets, and then must design marketing mixes to create these planned positions.

Positioning Maps

       d.   Perceptual positioning maps show consumer perceptions of their brands versus competing products on important buying dimensions. Figure 6-3 shows a positioning map for the U.S. large luxury sport utility vehicle market.

Choosing a Positioning Strategy

       e.   Each firm must differentiate its offer by building a unique bundle of benefits that appeals to a substantial group within the segment.

f.   The positioning task has three steps: identifying a set of possible competitive advantages; choosing the right competitive advantages; and selecting an overall positioning strategy. The company then needs to communicate and deliver the chosen position to the market.

1.   Positioning begins with actually differentiating the company's marketing offer so that it will give consumers more value than competitors' offers do. A company or market offer can be differentiated by product, services, channels, people, or image.

   i.   Product differentiation takes place along a continuum. At one extreme are products that vary little, while at the other extreme they are highly differentiated on features, performance, or style and design.

   ii.   Services differentiation can be done through speedy, convenient, or careful delivery. Installation can also differentiate a company, as can repair services. Other possibilities include training service or consulting services.

   iii. Channel differentiation can help a company gain competitive advantage through coverage, expertise, and performance.

   iv.   A company can differentiate on people—hiring and training people better than their competitors do.

   v.   A company can also differentiate on image. The chosen symbols, characters, and other image elements must be communicated through advertising that conveys the company's or brand's personality.

2.   The company must decide how many differences to promote and which ones.

   i.   Some marketers believe that the best strategy is to promote only one unique advantage, which can be called the unique selling proposition.

   ii.   Other marketers believe that companies should position themselves on more than one attribute, particularly if one or more companies are claiming to be best on the same attribute.

   iii. Not all brand preferences are meaningful or worthwhile. The company must carefully select which differences are worth promoting.

      a.   It should be important and deliver a highly valued benefit to target buyers.

      b.   It should be distinctive such that competitors do not offer the difference.

      c.   It should be superior, so that consumers cannot obtain the benefit elsewhere.

      d.   It should be communicable and visible to buyers.

      e.   It should be preemptive, so that competitors cannot easily copy it.

   f. It should be affordable.

   g. It should be profitable.

 iv. Consumers typically choose products and services that give them the greatest value. The full positioning of a brand is called the brand's value proposition—the full mix of benefits upon which the brand is positioned.

 v. Figure 6-4 shows possible value propositions with which a company might position its products.

   a. "More for more" positioning involves providing the most upscale product or service and charging a higher price to cover the higher costs.

   b. "More for the same" positioning introduces a brand offering comparable quality but at a lower price.

   c. "The same for less" offers good deals to customers.

   d. "Less for much less" positioning involves meeting consumers' lower performance or quality requirements at a much lower price.

   e. "More for less" is often claimed by companies, and in the short run, companies can often make this work. But in the long run, offering more usually costs more, so it is difficult to deliver on this promise.

 vi. Company and brand positioning should be summed up in a positioning statement. This statement should follow the form of *"To (target segment and need) our (brand) is (concept) that (point of difference)."* The statement first puts the product in a category and then shows the point of difference from other members in the category.

Communicating and Delivering the Chosen Position

g. When a company has chosen a position, it must take strong steps to deliver and communicate the desired position to target consumers.

h. All the company's marketing mix efforts must support the positioning strategy.

i. Designing the marketing mix—product, place, price, and promotion—involves working out the tactical details of the positioning strategy.

j. Companies often find it easier to come up with a good positioning strategy than to implement it. Establishing a position or changing one usually takes a long time. However, positions that took years to build can be lost easily. A company must take care to maintain the position through consistent performance and communication.

<u>Creative Marketing Exercises Designed to Reinforce the Concepts!!! (Suggested answers to these exercises can be found at the end of the Study Guide.)</u>

1. Identify the various segments used in the pasta industry. How can these segments become narrower? Explain.
2. Flip through your favorite magazine and locate products targeted to teens. How did you know they were the intended target market?
3. Evaluate your calendar and make a list of all the birthdays you celebrate in a year. What financial impact do these occasions have on your budget?
4. Go to www.sheraton.com and name the various segments targeted.
5. How many brands of diet beverages can you name? Is there a gap in this market? If so, where?
6. Analyze the current toothbrush market. Who are the main competitors and why do you think this market is so competitive?
7. Using your favorite search engine, identify an organization that impresses you with its social responsibility. What attributes attracted you to this organization? Explain.
8. You are the owner of a new jewelry store in your local shopping mall. What is your competitive advantage? Outline your plan.
9. Identify 5 examples of companies that market their product to the individual. What makes this approach work? Explain.
10. Which value proposition appeals to you the most and why? More for more, more for the same, same for less, less for much less, or more for less.

<u>"Linking the Concepts" (#1) – Suggestions/Hints</u>

1. One interesting example of a company using segmentation to market its products or services to customers would be Clemson University. Clemson, like other colleges and universities, is trying to get more people interested in its sports programs. Football is big at Clemson University and is popular with the males. A few years ago, Coach Tommy Bowden started a program that was designed to introduce and teach the basics of college football to women only. This course is open only to women, during the summer prior to the start of each football season. It is designed to get them involved and like football so they will come and enjoy the games, not just tailgate. It has become very popular. Other colleges and universities do similar programs for women and their sports programs.

2. Segments and descriptions of possible segments of the U.S. footwear market.

    a. Men's Shoes
       1. Executive wear – designed to wear at work and formal gatherings
       2. Sports wear – designed to be worn during sporting events. This can be broken down to types of sports (basketball, tennis, running, football, cross training, etc.)

3. Comfort wear – designed to be comfortable while walking and just wearing during a normal day
4. Name brand shoes – designed to appeal to the upper income or image concerned
5. Boots – this could be divided into work and style
6. Sandals, flip flops – this would be aimed at beach location users or summer casual, or camping/shower wear
7. Inexpensive wear – these shoes are designed for the price conscious men that want practical, yet not costly, footwear

b. Women's Shoes
1. They would be divided into similar segments as stated as above
2. Prom/formal wear – these shoes are designed for a prom, wedding, or some formal outing

c. Kid's Shoes
1. Sports wear – designed to participate in sports programs, such as football, basketball, soccer, etc.
2. Rain gear – designed to keep children from getting their feet wet
3. Boots – aimed at kids who go outdoors (camping, hiking)
4. Sandals, flip flops – this could be aimed at kids going on vacation to the beach, live at the beach, showers at camp, or summer casual
5. Cartoon character shoes – these shoes have their favorite cartoon characters and kids will like them just because they are their favorite TV/cartoon characters
6. Velcro strapped shoes – for ease of dressing or have not learned to tie their shoes
7. Dress shoes – for all of the times the kids have to dress up for occasions, holidays, school, or other important events

"Linking the Concepts" (#2) – Suggestions/Hints

1. Two examples that use different marketing strategies to segmenting markets are Sears and Footlocker. Sears uses more of a differentiated marketing strategy when it comes to selling its shoes. The company tries to offer an assortment of shoes that will appeal to a number of segments. There are shoes and boots designed for the work environment, for both men and women. There are formal styles for men and women. A consumer can purchase sporting shoes; however, there is not as great a variety that Footlocker might offer. There are men's, women's, and kid's shoes sold at Sears. Their prices range from shoes on sale (inexpensive) to name brand (Bass) shoes sold.

Footlocker, on the other hand, only offers sports-related shoes to their consumers. A consumer can find inexpensive to expensive sports shoes for just about every sport possible. There are basketball shoes, cleats, tennis shoes, running shoes as examples. Their product offering is for men, women, and kids. A consumer will not find formal or prom wear shoes at their stores.

2. Sears markets its image and shoes as simply a part of the total product mix offering available to consumers. Shoes do not represent the primary product offering of their store. Their image is quality products at reasonable prices for a variety of consumer needs. These needs include shoes, clothes, tools, washers and dryers. Consumers know, through experience and marketing efforts, that Sears offers a good variety of products and services. They have been successful in marketing this image. Their customer base is usually an older, married audience.

Footlocker markets itself as being on the edge with competitive and stylish products. They offer the top of the line running, basketball, and other sports shoes. They even show their products being worn by top professional athletes with posters and in-store TVs showing videos. The music in the stores is very trendy, upbeat, and modern. The sales force is usually a younger aged person. The colors in the store and other visual marketing are very bright, youthful, and energetic. Footlocker has been very successful in marketing itself with this image due to the fact that their customers are usually a younger audience as compared to Sears.

Marketing Adventure Exercises (Suggested answers to these exercises can be found at the end of the Study Guide.)

(Visit www.prenhall.com/adventure for advertisements.)

1.   Cosmetics   Edge

What are the three major steps in the target marketing process? Discuss each step as it relates to the selected ad.

2.   Financial   Bankmontreal, Pocketcard, Stroebrand, Zions

What are the various ways to segment the consumer market? Match the selected ads in the Financial category to each of these types of segmentation.

3.   Financial   Visa

Discuss the items included in demographic segmentation and why this is the most popular way to segment. Review the Visa ad in the Financial category and identify the multiple demographics variables used to segment the market.

4.   Exhibits   Student choice

Explain psychographic segmentation and find an ad in the Exhibits section that exemplifies both demographic and psychographic variables.

5.   Student choice

Consumers can be segmented based on behavior, occasion, benefits, user status, and loyalty status. Describe these and select an ad for each. Explain your reasons.

6.   Exhibits   Banrisul

Study the selected ad. Identify all the types of segmentation you see. Do you believe the use of multiple segmentation is effective?

7.   Internet   Hardwarecom

Compare and contrast segmenting the business market and the consumer market. Discuss the ways the business market can be segmented. Relate that information to the selected ad.

8.   Food   Student choice

Find ads in the Food category that are examples of intermarket segmentation. Discuss why this type of segmentation can be beneficial to marketers.

9.   Cosmetics   Student choice

Discuss the differences between differentiated and undifferentiated marketing. Find an ad for each in the Cosmetics section.

10.   Auto   Cadillac, Cadillac2

After segmenting and targeting, a company must decide what position it wants to occupy. Discuss positioning and identify the product positioning in the selected ads. Comment on the differences they each promote.

SUM IT UP!!!!!!

Using only this page, sum up all of the concepts and terms discussed in Chapter 6 – "Segmentation, Targeting, and Positioning: Building the Right Relationships with the Right Customers". Here is your chance to make sure you know and understand the concepts!!!!

Chapter 7
Product, Services, and Branding Strategy

Previewing the Concepts—Chapter Objectives

1. Define *product* and the major classifications of products and services.
2. Describe the decisions companies make regarding their individual products and services, product lines, and product mixes.
3. Discuss branding strategy----the decisions companies make in building and managing their brands.
4. Identify the four characteristics that affect the marketing of a service and the additional marketing considerations that services require.
5. Discuss two additional product issues: socially responsible product decisions and international product and services marketing.

JUST THE BASICS

Chapter Overview

In many ways, this chapter provides the information required to truly understand marketing. It focuses on the definition of what products and services are, and it provides details about branding.

After defining what a product is, the chapter goes on to detail the necessary attributes of products and services, as well as the branding, packaging, labeling, and product support decisions that marketers must make. There is information regarding product line and product mix decisions, and how to effectively manage both.

The section on branding provides a description of brand equity and the steps a company can take to build strong brands. Brand decisions such as positioning, name selection, sponsorship, and brand development are illustrated through use of examples.

Services marketing is differentiated from product marketing in that services are intangible, inseparable from the provider, highly variable, and perishable. Because of this, services marketers face additional challenges that product marketers do not. The service-profit chain, which links service firm profits with employee and customer satisfaction, has five key links that include internal service quality; satisfied and productive service employees; greater service value; satisfied and loyal customers; and healthy service profits and growth.

Finally, the social issues that affect product decisions are detailed, as well as the requirements for international product and services marketing.

Chapter Outline

1. **Introduction**
   a. Everything about FIJI Water contributes to a "Taste of Paradise" brand experience---from its name, packaging, and label to the places that sell and serve it, to the celebrities that endorse it. Skillful marketing people build the brand's ultra-chic image.
   b. Could any bottled water be worth $10 a bottle? Apparently so! FIJI is scrambling to keep up with surging demand. More and more people are buying into FIJI's "Taste of Paradise" brand promise, despite the high price---or maybe because of it. Clearly, water is more that just water when FIJI sells it.
   c. This chapter looks at the question *What is a product?* and then classifies products into consumer and business markets.

2. **What is a product?**
   a. A product is defined as anything that can be offered to a market for attention, acquisition, use, or consumption that might satisfy a need or a want. Broadly defined, products include physical objects, services, events, persons, places, organizations, ideas, or mixes of these entities.
   b. Services are a form of product that consist of activities, benefits, or satisfactions offered for sale that are essentially intangible and do not result in the ownership of anything.

   Products, Services, and Experiences
   c. Product is a key element in the market offering. Marketing-mix planning begins with formulating an offering that brings value to target customers and satisfies their needs.
   d. A company's marketing offer can provide both tangible goods and services. At one extreme, the offer may consist of a pure tangible good, while at the other extreme are pure services. Many goods and services combinations are available between these two extremes.
   e. Many companies are looking to deliver memorable experiences to differentiate their products and services. Whereas products and services are external, experiences are personal and take place in the minds of individual consumers. Companies that market experiences realize that consumers are really buying what the offers will do for them, not just the products and services themselves.

   Levels of Products and Services
   f. Products and services should be thought of on three levels (see Figure 7-1). Each level adds more customer value.
      1. The most basic level is the core benefit, which addresses what the consumer is really buying. It defines the core, problem-solving benefits or services that consumers seek.

2.   The second level is where the core benefit is turned into an actual product. The product's actual features, design, quality level, brand name, and packaging are developed.

3.   The third level is the augmented product, which brings in additional consumer services and benefits around the core benefits and actual product.

Product and Service Classifications

g.   Products and services fall into two broad classes based on the types of consumers that use them: consumer products and industrial products.

h.   Consumer products are bought by final consumers for personal consumption; they are generally classified by how consumers go about buying them.

1.   Convenience products are consumer products and services that the customer usually buys frequently, immediately, and with a minimum of comparison and buying effort. Convenience products are generally low priced, and marketers place them in many locations to make them readily available when customers need them.

2.   Shopping products are less frequently purchased. Customers carefully compare them on suitability, quality, price, and style. Consumers spend much more time and effort in gathering information and making comparisons. Shopping products are usually distributed through fewer outlets, but marketers provide deeper sales support to help customers in their comparison efforts.

3.   Specialty products have unique characteristics or brand identification for which a significant group of buyers is willing to make a special purchase effort. Buyers do not normally compare specialty products. They invest the time needed to reach dealers carrying these products, but no more.

4.   Unsought products are consumer products that the consumer does not know about or knows about but does not normally think of buying. Most major innovations are unsought until the consumer becomes aware of them, but the classic example of this type of product is insurance. Unsought products require a lot of advertising, personal selling, and other marketing efforts.

i.   Industrial products are those purchased for further processing or for use in conducting a business. There are three groups of industrial products and services.

1.   Materials and parts include raw materials and manufactured materials and parts.

2.   Capital items are industrial products that aid in a buyer's production or operations, including installations and accessory equipment.

3.   Supplies and services include operating supplies and repair and maintenance items. These are generally considered the convenience products of the industrial field.

j.   Organizations also carry out activities to sell the organization itself. Organization marketing consists of activities undertaken to create, maintain, or change the attitudes and behavior of target consumers toward an organization. Both profit and not-for-profit organizations market themselves. Corporate image advertising is a major tool companies use to market themselves to various publics.

k.   Person marketing consists of activities undertaken to create, maintain, or change attitudes or behavior toward particular people.

l.   Place marketing involves activities undertaken to create, maintain, or change attitudes or behavior toward particular places.

m.   Ideas can also be marketed. This area has been called social marketing, which is defined as the use of commercial marketing concepts and tools in programs designed to influence individuals' behavior to improve their well-being and that of society.

3.   **Product and Service Decisions**

a.   There are three levels of decision making for products and services: individual decisions, product line decisions, and product mix decisions.

Individual Product and Service Decisions

b.   Product benefits are communicated and delivered by product attributes such as quality, features, and style and design.

1.   Product quality is one of the marketer's major positioning tools. In the narrowest sense, *quality* can be defined as "freedom from defects," but most companies define quality in terms of customer satisfaction.

i.   Total quality management (TQM) is an approach in which all the company's people are involved in constantly improving the quality of products, services, and business processes. This approach has recently drawn some criticism, because too many companies viewed TQM as a cure-all and created token total quality programs that applied the principles superficially.

ii.   Today, many companies are using a "return on quality" approach, viewing quality as an investment and holding quality efforts accountable for bottom-line results.

iii.   Product quality has two dimensions: level and consistency. The quality level means performance quality or the ability of a product to perform its functions. Quality conformance means, quality consistency, freedom from defects and consistency in delivering a targeted level of performance

2.   A product can be offered with varying features. Features are a competitive tool for differentiating the company's product from competitors' products.

i.   The company should periodically survey buyers who have used the product to ask *How do you like the product?*.

Which specific features of the product do you like most? Which features could we add to improve the product? The company can then assess each feature's value to customers versus its cost to the company.

3.   A way to add value is through distinctive product style and design.

    i.   *Style* describes the appearance of a product.

    ii.   *Design* goes to the heart of a product. Good design contributes to a product's usefulness as well as its looks.

    iii.   Good style and design can attract attention, improve product performance, cut production costs, and give the product a strong competitive advantage.

4.   A brand is a name, term, sign, symbol, or design, or a combination of these, that identifies the maker or seller of a product or service. Consumers view brands as an important part of the product.

    i.   Branding helps buyers by identifying products that might help them, and it also tells them something about product quality.

    ii.   Branding helps sellers also. The brand name becomes the basis on which a whole story can be built about a product's special qualities. The brand name and trademark can provide legal protection for unique product features that otherwise might be copied by competitors.

5.   Packaging involves designing and producing the container or wrapper for a product. The package includes a product's primary container, and may include a secondary package that is thrown away when the product is about to be used. There can also be a shipping package, and labeling is also part of packaging.

    i.   Many factors have made packaging an important marketing tool. Clutter on retail shelves means that packages must now perform sales tasks such as attracting attention, describing the product, and making the sale.

    ii.   Poorly designed packages can create problems for consumers and lost sales for the company.

    iii.   The packaging concept states what the package should be or do for the product. Then decisions need to be made on specific elements of the package, such as size, shape, materials, color, text, and brand mark.

    iv.   Product safety has also become a major packaging concern. Many companies have also reduced their packaging and begun using environmentally responsible materials.

6.   Labels can range from simple tags to complex graphics that are part of the package. Labels identify the product or brand; they could describe several things about the product; and they might promote the product through attractive graphics.

    i.   Several federal and state laws regulate labeling. One act requires that labels include unit pricing, open dating, and nutritional labeling. Others set mandatory labeling requirements and allow federal agencies to set packaging regulations in specific industries.

7. Customer service is another element of product strategy. A company usually includes some support services in its offer.

    i. Again, the first step is to survey customers periodically to assess the value of current services and to obtain ideas for new ones.

    ii. The company then has to assess the cost of providing these services.

    iii. Many companies are using a mix of phone, email, fax, Internet, and interactive voice and data technologies to provide support services.

Product Line Decisions

c. A product line is a group of products that are closely related because they function in a similar manner, are sold to the same customer groups, are marketed through the same types of outlets, or fall within given price ranges.

d. The major product line decisions involve product line length, which is the number of items in the product line.

    1. The line is too short if the manager can increase profits by adding items. The line is too long if the manager can increase profits by dropping items.

    2. The length of the product line is influenced by the company's objectives and resources.

    3. A company can lengthen its product line by either line stretching or by line filling.

        i. Line stretching occurs when a company lengthens its product line beyond its current range. The line can be stretched downward, upward, or both ways.

        ii. Product line filling is the process of adding more items within the present range of the line.

Product Mix Decisions

e. A product mix (or product assortment) consists of all the product lines and items that a particular seller offers for sale.

f. A company's product mix has four important dimensions: width, length, depth, and consistency.

    1. Product mix width refers to the number of different product lines the company carries.

    2. Product mix length refers to the total number of items the company carries within its product lines.

    3. Product mix depth refers to the number of versions offered of each product in the line.

4. Product mix consistency refers to how closely related the various product lines are in end use, production requirements, distribution channels, or some other way.

g. The company can increase its business in four ways. It can add new product lines, widening its product mix. It can lengthen its existing product lines to become a more full-line company. It can add more versions of each product and deepen its product mix. Or it can pursue more product line consistency—or less—depending on whether it wants to have a strong reputation in a single field or in several fields.

4. **Branding Strategy: Building Strong Brands**

a. Some analysts see branding as *the* major enduring asset of a company, outlasting the company's specific products and facilities. Thus, brands are powerful assets that must be carefully developed and managed.

Brand Equity

b. Brands represent consumers' perceptions and feelings about a product and its performance—everything that the product or service means to consumers. Brands exist in the minds of consumers.

c. Brand equity is the positive differential effect that knowing the brand name has on customer response to a product or service. A measure of a brand's equity is the extent to which customers are willing to pay more for the brand.

d. A brand with strong brand equity is a valuable asset. Brand valuation is the process of estimating the total financial value of a brand. Measuring value is difficult.

e. A powerful brand enjoys a high level of consumer brand awareness and loyalty. Because consumers expect stores to carry the brand, the company has more leverage in bargaining with resellers.

f. A powerful brand forms the basis for building strong and profitable customer relationships. The fundamental asset underlying brand equity is customer equity—the value of the customer relationships that the brand creates. What a powerful brand represents is a set of loyal customers.

Building Strong Brands

g. Figure 7-3 shows that the major brand strategy decisions involve brand positioning, brand name selection, brand sponsorship, and brand development.

h. Marketers need to position their brands clearly in target customers' minds. You can position brands at any of three levels.

1. The lowest level is positioning the brand on product attributes. But competitors can easily copy attributes, and customers aren't interested in attributes as such; they are interested in what the attributes will do for them.

2. A brand can be positioned by associating its name with a desirable benefit.

3. The strongest brands are positioned on strong beliefs and values. These brands pack an emotional wallop.

i. When positioning a brand, the marketer should establish a mission for the brand and a vision of what that brand must be and do. A brand is the company's promise to deliver a specific set of features, benefits, services, and experiences consistently to the buyers.

j. A good brand name adds greatly to a product's success. Desirable qualities for a brand name include the following:
   1. It should suggest something about the product's benefits and qualities.
   2. It should be easy to pronounce, recognize, and remember.
   3. It should be distinctive.
   4. It should be extendable.
   5. It should translate easily into foreign languages.
   6. It should be capable of registration and legal protection.

k. A manufacturer has four sponsorship options, including: launching a manufacturer's or national brand; selling to a reseller who gives it a private brand (also called store brand or distributor brand); licensing a brand; or joining forces with another company and co-brand.
   1. Manufacturer's brands have long dominated in retail, but an increasing number of retailers and wholesalers have created their own private brands.
      i. Private brands can be hard to establish and costly to stock and promote, however they also yield higher profit margins for the retailer.
      ii. Taken as a single brand, private-label products are the number one, two, or three brand in more than 40% of all grocery product categories. They capture more than a 20% share of sales in U.S. supermarkets, drug chains, and mass merchandise stores. Private-label apparel captures a 36% share of all U.S. apparel sales.

      iii. In the battle of the brands between manufacturers' and private brands, retailers have many advantages. Most retailers charge manufacturers' slotting fees, which are payments demanded by retailers before they will accept new products and find "slots" for them on the shelves.

      iv. To fend of private brands, leading brand marketers will have to invest in R&D to bring out new brands, new features, and continuous quality improvements. They must design advertising programs to maintain high awareness and find ways to partner with major distributors.

2. Some companies license names or symbols previously created by other manufacturers, names of well-known celebrities, or characters from popular movies and books.

    i. Name and character licensing has grown rapidly. Annual retail sales of licensed products in the United States and Canada have grown from only $4 billion in 1977 to $55 billion in 1987 and more than $105 billion today.

3. Co-branding occurs when two established brands of different companies are used on the same product. In most co-branding situations, one company licenses another company's well-known brand to use in combination with its own.

    i. Co-branding has many advantages. The combined brands create broader consumer appeal and greater brand equity. Co-branding allows a company to expand its existing brand into a category it might otherwise have difficulty entering alone.

    ii. Co-branding also has limitations, which usually involve complex legal contracts and licenses. Co-branding partners must carefully coordinate their advertising, sales promotion, and other marketing efforts. Each partner must trust the other will take good care of its brand.

4. A company has four choices when it comes to developing brands (see Figure 7-4). It can introduce line extensions, brand extensions, multibrands, or new brands.

    i. Line extensions occur when a company introduces additional items in a given product category under the same brand name, such as new flavors, forms, colors, ingredients, or package sizes.

        a. The vast majority of all new-product activity consists of line extensions.

        b. A company could introduce line extensions as a low-cost, low-risk way to introduce new products. Or it might want to meet consumer needs for variety, to use excess capacity, or simply to command more shelf space from resellers.

        c. Line extensions involve some risks. An over-extended brand name might lose its specific meaning, or heavily extended brands can cause customer confusion or frustration. Sales of an extension could come at the expense of other items in the line.

    ii. A brand extension involves the use of a successful brand name to launch new or modified products in a new category.

        a. A brand extension gives a new product instant recognition and faster acceptance.

b. But the extension may confuse the image of the main brand. If a brand extension fails, it may harm attitudes toward the other products carrying the same brand name.

iii. Multibranding offers a way to establish different features and appeal to different buying motives.

a. A major drawback of multibranding is that each brand might obtain only a small market share, and none may be very profitable. The company may end up spreading its resources over many brands instead of building a few brands to a highly profitable level.

iv. New brands can be created when a company believes that the power of its existing brand name is waning, thus a new one is needed. Or a company can create a new brand name when it enters a new product category for which none of the company's current brand names is appropriate.

a. Offering too many brands can result in a company spreading its resources too thin.

Managing Brands

l. Companies must carefully manage their brands.

1. Customers come to know a brand through a wide range of contacts and touchpoints. These include advertising, but also personal experience with the brand, word of mouth, personal interactions with company people, telephone interactions, company web pages, and many others. Any of these experiences can have a positive or negative impact on brand perceptions and feelings.

2. The brand's positioning will not take hold fully unless everyone *lives* the brand. Companies carry on internal brand building to help employees understand, desire, and deliver on the brand promise.

m. Brand managers do not have enough power or scope to do all the things necessary to build and enhance their brands.

1. Many companies are setting up brand asset management teams to manage their major brands. Companies have also appointed brand equity managers to maintain and protect their brands' images, associations, and quality, and to prevent short-term actions by overeager brand managers from hurting the brand.

n. Companies need to periodically audit their brands' strengths and weaknesses. The brand audit may turn up brands that need to be repositioned because of changing customer preferences or new competitors. Some cases may call for a complete "rebranding" of a product, service, or company.

5. **Services Marketing**

a. Service now account for 72.5 percent of U.S. gross domestic product and nearly 60 percent of personal consumption expenditures. Between 2002 and 2012, an estimated 96 percent of all new jobs generated in the U.S.

will be in service industries. Services are growing even faster in the world economy, making up 20 percent of the value of all international trade.

b.   Service industries include governments, private not-for-profit organizations, and businesses that offer services.

Nature and Characteristics of a Service

c.   A company must consider four special service characteristics when designing marketing programs: intangibility, inseparability, variability, and perishability. These characteristics are outlined in Figure 7-5.

1.   Service intangibility means that services cannot be seen, tasted, felt, heard, or smelled before they are bought. To reduce uncertainty, buyers look for "signals" of service quality, drawing conclusions from the place, people, price, equipment, and communications that they can see.

2.   Service inseparability means that services cannot be separated from their providers, whether the providers are people or machines. Because the customer is also present as the service is produced, provider-customer interaction is a special feature of services marketing.

3.   Service variability means that the quality of services depends on who provides them as well as when, where, and how they are provided.

4.   Service perishability means that services cannot be stored for later sale or use.

Marketing Strategies for Service Firms

d.   In a service business, the customer and front-line service employees interact to create the service. Thus, service providers must interact effectively with customers to create superior value during service encounters.

1.   The service-profit chain links service firm profits with employee and customer satisfaction. This chain consists of five links:
    i.   Internal service quality.
    ii.   Satisfied and productive service employees.
    iii.   Greater service value.
    iv.   Satisfied and loyal customers.
    v.   Healthy service profits and growth.

2.   Figure 7-6 shows that service marketing also requires internal marketing and interactive marketing.
    i.   Internal marketing means that the service firm must effectively train and motivate its customer-contact employees and supporting service people to work as a team to provide customer satisfaction. Internal marketing precedes external marketing.

91

  ii. Interactive marketing means that service quality depends heavily on the quality of the buyer-seller interaction during the service encounter.

3. The solution to price competition is to develop a differentiated offer, delivery, and image.

  i. The offer can include innovative features that set one company's offer apart from competitors' offers.

  ii. Service companies can differentiate their service delivery by having more able and reliable customer-contact people, by developing a superior physical environment in which the service is delivered, or by designing a superior delivery process.

  iii. Service companies can work on differentiating their images through symbols and branding.

4. One of the major ways a service firm can differentiate itself is by delivering consistently higher quality than its competitors do.

  i. Service quality is harder to define and judge than is product quality. Customer retention is probably the best measure of quality—a service firm's ability to hang onto its customers depends on how consistently it delivers value to them.

  ii. Service quality will always vary, depending on the interactions among employees and customers.

  iii. Good service recovery can turn angry customers into loyal ones. In fact, good recovery can win more customer purchasing and loyalty than if things had gone well in the first place.

  iv. The first step is to empower front-line employees---to give them authority.

5. Service firms are under great pressure to increase service productivity.

  i. They can do this by training current employees better or by hiring new ones who will work harder or more skillfully.

  ii. Companies must avoid pushing productivity so hard that doing so reduces quality.

6. Additional Product Considerations

<u>Product Decisions and Social Responsibility</u>

a. The government may prevent companies from adding products through acquisitions if the effect threatens to lessen competition.

b. Companies dropping products must be aware that they have legal obligations to their suppliers, dealers, and customers who have a stake in the dropped product.

c. Manufacturers must comply with specific laws regarding product quality and safety. Product liability suits are now occurring in federal and state courts at the rate of almost 110,000 per year, with a median jury award of

$6 million and individual awards often running into the tens or even hundreds of millions of dollars.

International Product and Services Marketing

d.　International product and service marketers face special challenges. On the one hand, companies would like to standardize their offerings. This helps a company to develop a consistent worldwide image and lowers manufacturing costs and eliminates duplication of research and development, advertising, and product design efforts.

e.　On the other hand, consumers around the world differ in their cultures, attitudes, and buying behaviors. And markets vary in their economic conditions, competition, legal requirements, and physical environments. Companies must usually respond to these differences by adapting their product offerings.

f.　Packaging presents new challenges for international marketers. Names, labels, and colors may not translate easily from one country to another. Packaging may have to be tailored to meet the physical characteristics of consumers of various parts of the world.

g.　Service marketers face special challenges when going global. Some service industries have a long history of international operations. Professional and business service industries such as accounting, management consulting and advertising have only recently globalized. Retailers are among the latest service businesses to go global.

h.　Despite such difficulties, the trend toward growth of global service companies will continue.

Creative Marketing Exercises Designed to Reinforce the Concepts!!!　(Suggested answers to these exercises can be found at the end of the Study Guide.)

1.　Call your local physician and ask what products they sell versus what products they offer.

2.　Outline the "product" your college/university is selling.

3.　Go to www.esteelauder.com and look at their skin care line.　Identify five products that can be viewed as unsought.

4.　Go to www.amazon.com and read a review of the book by Private Jessica Lynch.　How is this book an example of "person marketing?"

5.　Visit www.georgia.org and determine how Georgia is positioning the state as a tourism destination.

6.　Go to www.bestbuy.com and compare the features on two brands of DVD players and determine which player is the better value.

7.　Visit www.marykay.com and locate their "TimeWise" line of anti-aging products.　How is this line an example of product expansion?

8.　Visit www.polo.com and outline how Ralph Lauren has kept his product lines consistent with his mission.

9. Visit www.uspto.gov and conduct a search for "NASCAR." How many trademarks have been issued to this organization?

10. During your next visit to your local grocery store, locate a product that you think is a good example of brand extension. Justify your selection.

"Linking the Concepts" – Suggestions/Hints

1. From Procter and Gamble's Web site (www.pg.com), the following are the product lines and individual products offered:

    a. Laundry and Cleaning Products
       1. Bold
       2. Bounce
       3. Cascade
       4. Cheer
       5. Comet
       6. Joy
       7. Dawn
       8. Ivory Dish
       9. Downy
       10. Dreft
       11. Mr. Clean
       12. Spic and Span
       13. Tide
       14. Era
       15. Gain
       16. Ivory Snow
       17. Oxydol

    b. Health Care Products
       1. Living Better
       2. Metamucil
       3. Pepto Bismol
       4. Attends
       5. Crest
       6. Fixodent
       7. Gleem
       8. Scope
       9. Vicks Vapo Rub
       10. Chloraseptic
       11. DayQuil
       12. NyQuil
       13. Sinex
       14. Vicks 44

    c. Beauty Products
        1. Cover Girl
        2. Max Factor
        3. Old Spice
        4. Sure
        5. Giorgio Beverly Hills
        6. Hugo Boss
        7. Laura Bisgiioti-roma
        8. Red
        9. Venezia
       10. Wings
       11. Head and Shoulders
       12. Ivory Hair Care
       13. Pantene
       14. Pert
       15. Rejoy-Rejoice
       16. Vidal Sassoon
       17. Camay
       18. Clearisil
       19. Coast
       20. Ivory Soap
       21. Muse
       22. Oil of Olay
       23. Safeguard
       24. Zest

    d. Food and Beverage Products
        1. Crisco
        2. Folgers
        3. Jif
        4. Pringles
        5. Sunny Delight
        6. Millstone
        7. Olean
        8. Tender Leaf Tea

The last group (Food and Beverage Products) seems to be the most surprising of the product lines. All of the other products seem to be related by cleanliness and health. The food product line contains Crisco. This is not a "health" item, as compared to the rest of the products offered.

2.   Procter and Gamble's product mix is relatively consistent. All of the items are related by the fact that they are products that could be used for cleaning or for health purposes (except Crisco). The food items are all, relatively, healthy. They include coffee, tea, juice, and peanut butter. The overall strategy of the company seems to be the offering of items that consumers will use in the

operations of a quality household. This household product offering strategy seems very logical and it allows for easy product extensions as consumers' home and health care needs change.

Marketing Adventure Exercises (Suggested answers to these exercises can be found at the end of the Study Guide.)

(Visit www.prenhall.com/adventure for advertisements.)

1. Student choice

A product is defined as anything that can be offered to a market for attention, acquisition, use, or consumption, and that might satisfy a want or need. Since products are more than just tangible goods, list all the other components of product and find an ad to match as many as possible.

2. Food   Student choice

The company's marketing offering may be a product that is both a tangible good and a service. Find an ad in the Food category that is a combination of both good and service.

3. Auto   Cadillac

What are the three levels of a product as they relate to the selected ad?

4. Student choice

Discuss the four ways that consumer products can be classified. Select ads for each of the classifications.

5. Electronics   Axiom, Samsung

Discuss the differences between industrial products and consumer products. Would the products in the selected ads be consumer or industrial products?

6. Financial   MYOB
     Internet   Hardwarecom
     Services   Corts

List and describe the three groups of industrial products. Match the selected products with their appropriate group.

7. Student choice

What is organizational marketing? Find an example from profit and not-for-profit organizations and discuss the purpose of the ad.

8.   Student choice

How does the Social Marketing Institute define the marketing of social ideas? Select those ads that you believe are examples of effective social marketing.

9.   Food   Student choice

What is brand positioning? Find an ad in the Food category and discuss its brand positioning.

SUM IT UP!!!!!!

Using only this page, sum up all of the concepts and terms discussed in Chapter 7 – "Product, Services, and Branding Strategy". Here is your chance to make sure you know and understand the concepts!!!!

Chapter 8
New-Product Development and Product Life-Cycle Strategies

Previewing the Concepts: Chapter Objectives

1.   Explain how companies find and develop new product ideas.
2.   List and define the steps in the new product development process.
3.   Describe the stages of the product life cycle.
4.   Describe how marketing strategies change during the product's life cycle.

JUST THE BASICS

Chapter Overview

For companies to succeed, they must be constantly developing new products and services. Even a company as well-known in the marketplace as Apple Computer must constantly innovate. Firms can either acquire new products through buying a whole company, a patent, or a license to produce a product, or through internal development in their own research and development department.

New product development should be a systematic process; this can lessen the probability of failure. New products fail at an alarming rate—some estimates are as high as 95% of all new products fail. To avoid that fate, companies follow a new product development process, which includes idea generation, idea screening, concept development and testing, marketing strategy development, business analysis, product development, test marketing, and commercialization. As each step is accomplished, a firm go/no go decision is made as to whether to continue. This is done because as each successive stage is passed, new product development costs increase. Therefore, it makes sense to cut the product development process short if it appears the product will not be a success in the marketplace.

All products follow a life cycle. Some go through the cycle quickly, others can take many years. The standard product life cycle (PLC) includes the phases of introduction, growth, maturity, and decline. The PLC concept can describe a product class, a product form, or a brand. Most products are in the mature stage of the PLC.

Chapter Outline

1.   **Introduction**
       a.   Apple Computer, an early product innovator, got off to a fast start, but only a decade later, as its creative fires cooled, Apple was on the brink of

extinction. This set the stage for one of the most remarkable turnarounds in corporate history.

b. By the mid- to late-nineties, Apple's sales had plunged 50 percent, yet by 2005 Apple had engineered a turnaround. Apple rediscovered the magic that had made the company so successful in the first place: dazzling creativity and customer-driven innovation.

c. The first task was to revitalize Apple's computer business, launching the iMac which won raves for design and lured buyers in droves. The Mac OS X was unleashed, a ground-breaking new operating system which served as the launching pad for a new generation of Apple computers and software products.

d. Then Apple introduced the iPod, a tiny computer with an amazing capacity to store digital music, on of the greatest consumer electronics hits of all time. By January 2005, Apple had sold 10 million iPods. As the Apple story suggests, a company has to be good at developing and managing new products. Every product goes through a product life cycle.

2. New Product Development Strategy

a. A firm can obtain new products in two ways. One is through acquisition of a whole company, a patent, or a license to produce someone else's product. The other way is through new product development in the company's own research and development department.

b. New products are original products, product improvements, product modifications, and new brands that the firm develops.

c. New product development is risky. One source estimates that no more than 10% of new products are still on the market and profitable after three years. Failure rates for new industrial products may be as high as 30%.

d. There are several reasons new products fail. Although an idea may be good, the market size may have been overestimated. The actual product might not have been designed as well as it should have been. The product could be incorrectly positioned in the market, priced too high, or poorly advertised. Sometimes the cost of development is higher than expected.

e. To lower these risks, companies need to set up a systematic new product development process for finding and growing new products. Figure 8-1 shows the eight major steps in this process.

Idea Generation

f. New product development begins with idea generation, which is a systematic search for new product ideas. A company normally has to generate many ideas in order to find a few good ones.

g. Major sources of new product ideas include both internal and external sources.

1. The company can use internal sources to find new ideas through formal research and development. It can get the ideas from executives, scientists, engineers, manufacturing staff, and sales-people.

2. Good external sources include watching and listening to customers. The company can analyze the questions and complaints of customers to find new products. The company can also have engineers or salespeople meet and work with customers to get suggestions. It can also conduct surveys or focus groups to learn about customer needs and wants. Consumers can often create new products and uses on their own, and sometimes companies can give consumers the resources to design their own products.

h. For some products, especially technical ones, customers may not know what they need.

i. Competitors can be a good source of new product ideas. Companies can watch competitors' ads to get clues about their products. They can purchase competing products, take them apart, analyze their sales, and decide if they should bring out a new product of their own.

j. Distributors and suppliers can also contribute new product ideas. Other sources include trade magazines, shows, and seminars, as well as government agencies, new product consultants, advertising agencies, marketing research firms, university and commercial laboratories, and individual inventors.

k. The search for new products should be systematic. Management can develop an idea management system that directs new ideas to a central point where they can be collected, reviewed, and evaluated.

Idea Screening

l. Idea generation is all about creating a large number of ideas. Idea screening is the first idea-reducing stage. Product development costs rise rapidly in later stages, so companies want to go ahead with only the product ideas that will become profitable.

Concept Development and Testing

m. An attractive idea should be developed into a product concept. A product concept is a detailed version of the idea stated in meaningful consumer terms. A product image is the way consumers perceive an actual or potential product.

1. In concept development, several descriptions of the product are generated to find out how attractive each concept is to customers. From these concepts, the best one is chosen.

2. Concept testing calls for testing new product concepts with groups of target consumers. The concepts may be presented symbolically or physically. Table 8-1 lists some questions that might be asked of consumers after they have been exposed to the product concept. The answers will help the company decide which concept has the strongest appeal.

101

Marketing Strategy Development

n.   Marketing strategy development entails designing an initial marketing strategy for introducing the product to the market. It consists of three parts:

    1.   The first part describes the target market, the planned product positioning, and the sales, market share, and profit goals for the first few years.

    2.   The second part outlines the product's planned price, distribution, and marketing budget for the first year.

    3.   The third part describes the long-run sales, profit goals, and marketing mix strategy.

Business Analysis

o.   Now the company evaluates the business attractiveness of the proposed product. Business analysis involves a review of the sales, costs, and profit projections for a new product to find out whether they satisfy the company's objectives.

    1.   To estimate sales, the company could look at the sales history of similar products, or conduct surveys of market opinion. It can then estimate the minimum and maximum sales to assess the range of risk.

Product Development

p.   If the product concept passes the business analysis stage, it moves into product development. Here, R&D or engineering develops the product concept into a physical product. There is now a large jump in investment, but this stage will show whether the product idea is workable.

q.   Often, products will undergo rigorous tests to make sure that they perform safely and effectively. Tests might also show if consumers will find value in them.

Test Marketing

r.   This is the stage in which the product and marketing program are introduced into more realistic market settings. Test marketing gives the marketer experience with marketing the product before going to the great expense of a full introduction.

s.   The amount of test marketing varies with each new product.

    1.   When introducing a new product requires a big investment, or when management is not sure of the product or marketing program, a company may do a lot of test marketing. Although these costs can be high, they are often small when compared to the costs of making a major mistake.

Commercialization

t.   Commercialization involves introducing the product into the market. The first decision that needs to be made is timing. The second decision involves where to launch the product—a single location, a region, the national market, or the international market. Many companies will develop a planned market rollout.

Organizing for New Product Development

u. Under sequential product development, one company department works individually to complete its stage of the new product development process before passing the new product along to the next department and stage.
    1. This orderly process can help bring control to complex and risky projects. But it can also be very slow.

v. The simultaneous product development (or team-based or collaborative product development) concept can be used to get products to market more quickly.
    1. Company departments work closely together through cross-functional teams, overlapping the steps in the product development process to save time and increase effectiveness.
    2. Teams usually include people from the marketing, financing, design, manufacturing, and legal departments. Sometimes even suppliers and customer companies are involved.
    3. There are some limitations to this approach. Fast product development can be riskier and more costly than the more orderly process. There can also be increased organizational tension and confusion. The company must also ensure that quality is not sacrificed by rushing a product to market.
    4. But the rewards outweigh the risks. Companies that get new and improved products to the market faster than competitors often gain competitive advantage.

3. **Product Life-Cycle Strategies**
    a. A typical product life cycle (PLC) is shown in Figure 8-2. It shows the course that a product's sales and profits take over its lifetime. There are five distinct PLC stages:
        1. Product development begins when the company finds and develops new product ideas. Sales are zero in this stage and the company's investment costs mount.
        2. Introduction is a period of slow sales growth as the product is introduced into the market. Profits are nonexistent because of the heavy expenses of product introduction.
        3. Growth is a period of rapid market acceptance and increasing profits.
        4. Maturity is a period of slower sales growth because the product has achieved acceptance by most potential buyers. Profits level off or decline because of increased marketing outlays.
        5. Decline is the period when sales fall off and profits drop.

b.   Not all products follow this product life cycle. Some products are introduced and die quickly, others stay in the mature stage for a very long time. Some can enter the decline stage and then cycle back into the growth stage through strong promotion or repositioning.

c.   The PLC concept can describe a product class, such as gasoline-powered automobiles; a product form, such as SUVs; or a brand, such as the Ford Explorer.

    1.   Product classes have the longest life cycles.

    2.   Product forms tend to have the standard PLC shape.

d.   The PLC concept can also be applied to styles, fashions, and fads. Their special life cycles are shown in Figure 8-3.

    1.   A style is a basic and distinctive mode of expression.

    2.   A fashion is a currently accepted or popular style in a given field.

    3.   Fads are fashions that enter quickly, are adopted with great zeal, peak early, and decline very quickly. They last only a short time and tend to attract only a limited following.

e.   Using the PLC concept for forecasting product performance or for developing marketing strategies creates some difficulties. Marketers could have trouble identifying which stage of the PLC the product is in, or in pinpointing when the product moves into the next stage. It can also be difficult to determine why the product moves through the stages. Using the PLC concept to development marketing strategy also can be difficult because strategy is both a cause and a result of the product's life cycle.

Introduction

f.   The introduction stage starts when the new product is first launched.

g.   In this stage, profits are negative or low because of low sales and high distribution and promotion expenses. Promotion spending is relatively high to inform consumers of the new product.

Growth Stage

h.   If the new product satisfies the market, it will enter a growth stage in which sales will begin climbing quickly. Early adopters will continue to buy, and later buyers will start following their lead.

i.   New competitors will enter the market. This leads to an increase in the number of distribution outlets. Prices remain where they are or fall slightly. Companies keep their promotion spending at the same or a slightly higher level.

j.   Profits increase during the growth stage as promotion costs are spread over a larger volume and as unit manufacturing costs fall.

Maturity Stage

k.   As the product's sales growth slows down, the product enters a maturity stage. This stage normally lasts longer than the previous stages. Most products are in the maturity stage of the life cycle, and therefore most of marketing management deals with the mature product.

l.   Competitors begin marking down prices while increasing their advertising and sales promotions. They may also up their R&D budgets to find better versions of the product. These steps lead to a drop in profit.

m.   Product managers should consider modifying the market, product, and marketing mix at this point.

1.   In modifying the market, companies try to increase the consumption of the current product. It looks for new users and market segments.

2.   The company may also try to modify the product by changing characteristics such as quality, features, or style to attract new users and to inspire more usage. Or the company could add new features to expand the product's usefulness, safety, or convenience.

3.   The company can modify the marketing mix to improve sales. It can cut prices, launch a better advertising campaign, or use aggressive sales promotions.

Decline Stage

n.   The sales of most product forms and brands eventually dip. This is the decline stage.

o.   Sales decline for many reasons, including technological advances, shifts in consumer tastes, and increased competition. Some firms will withdraw from the market; others may prune their product offerings.

p.   Marketers could decide to maintain the brand without change in the hope that competitors may leave the industry. Or management may decide to reposition or reformulate the brand in hopes of moving it back into the growth stage of the PLC. Marketers could also decide to harvest the product, which means reducing various costs while hoping that the sales hold up. Or the company could simply drop the product.

q.   Table 8-2 summarizes the key characteristics of each stage of the product life cycle. It also lists the marketing objectives and strategies for each stage.

Creative Marketing Exercises Designed to Reinforce the Concepts!!! (Suggested answers to these exercises can be found at the end of the Study Guide.)

1.   Visit www.spriteremix.com and determine in which stage of the product life cycle this product is currently.

2.   Conduct a brief research study of the lodging industry. Explain why there are so many smaller hotels being acquired by the larger hotels.

3.   Compile a list of 5 products that only lasted months on the open market. How would you explain their demise?

4.   Brainstorm ideas for new products with 2 friends and then again with 2 older adults. Is there a significant difference in concepts? Explain.

5.   Interview the manager of a local fast food restaurant and ask how the company develops ideas for new products.

6.   Identify 5 examples of print ads where the image and concepts seem mismatched to the product. Justify your selection.

7.   Go to www.lasvegas.com and find 5 examples of hotels/casinos that use virtual reality to promote their properties.

8.   Using your favorite search engine, find 5 websites that show prototypes for new products.

9.   Using your favorite magazines and websites, find examples of products that differentiate between a style, a fashion, and a fad.

10.   Find 3 websites that represent a company in the maturity phase. What are they doing to promote growth instead of decline?

"Linking the Concepts" (#1) -- Suggestions/Hints

1. One example of a product that was introduced recently would be cell phones without dialing pads. This would have to be the "Best New Product of the Year." Nokia designed a cell phone without a dialing pad. Instead, there is a mirror on the side of the cell phone that converts into a color LCD screen. This allows you to place your call by speech recognition (voice dialing or voice commands) or an I-Pod like rotary button. The cell phone has been discussed on the major news channels concerning their use. This product has created the most buzz worldwide.

   The development process was focused on the factor of convenience. As cell phones have evolved, this seemed to be the logical evolution...more features offering more convenience.

2. An idea generated for a new snack food might be chocolate covered Pringles. This product would be the chocolate covered chip stacked in a small package and sold as a snack.

   During the idea generation stage, this might seem like a good idea due to the fact that there is market demand, currently, for chocolate covered pretzels. This could be an extension of that idea. In the testing lab, the different flavors of currently marketed Pringles could be dipped in regular, dark, and white chocolate for testing.

   The marketing strategy could be to market this product to the same target market as the chocolate covered pretzel consumers market. The business analysis would have to work out the production issues and the placement issues. This would be a bulkier item to place on the shelves with other snack items.

   The product would have to be developed and tested in-house and by consumers in grocery stores to determine if the idea would sell. If successful, then the product would be introduced nationwide, possibly with a movie release and at the same target market as chocolate covered pretzel consumers.

"Linking the Concepts" (#2) -- Suggestions/Hints

1. The concept of today's razor is the same concept as the razors in the 1950's, except the blades are sharper and some razors have more of them. The shaving process is still the same – drag a sharp razor across the face to remove whiskers. Most improvements to the razor are to make the experience of shaving less painful and irritating. There are more razors stacking the number of blades for closer shaves and lotion strips have been added to smooth the process. Most of the changes to razors are examples of product improvements that are designed for products in the mature stage of the product life cycle.

2. Although the Crayon brand has been around since 1903, it continues to grow and remains one of the most recognizable brands in the consumer marketplace. Binney & Smith has worked hard to equate the brand with color, fun, quality, and development over the years since its inception. In addition, the Crayola product line has grown significantly over the years to not only include crayons, but to also include markers, colored pencils, paints, modeling compounds and craft and activity products. The company currently produces over 100 different crayon colors.

Marketing Adventure Exercises (Suggested answers to these exercises can be found at the end of the Study Guide.)

(Visit www.prenhall.com/adventure for advertisements.)

1.   Electronics   Playstation, Playstation2

New product ideas can come from both internal and external sources. Discuss possible sources of information for the selected ad.

2.   Cosmetics   Head and shoulders

Describe the three parts of the marketing strategy statement as they relate to the selected ad.

3.   Food   Burger King

Discuss test marketing and the specific risks it poses to a company like Burger King. Do you think Burger King test markets all its new offerings?

4.   Cosmetics   Colgate

Colgate is a company with international distribution systems. Your text states that Colgate used a "lead country" strategy in the past, and now it uses a swift global assault when introducing new products. Explain these two strategies for distributing new products in the global marketplace.

5.   Autos   Student choice

A product life cycle can describe a product class, a product form, or a brand. Relate this to the product in the selected ad. Discuss the time frame for each.

6.   Financial   Student choice

Describe the various activities during the introduction stage of the product life cycle. Select an ad from the Financial category that is appropriate for this stage.

7.   Financial   Student choice

What changes does a product go through in the growth stage? Select an ad that is appropriate for this stage. Explain.

8.   Financial   Student choice

Which stage of a product's life tends to be the longest? Select an ad that is representative of this stage and explain its purpose.

SUM IT UP!!!!!!

Using only this page, sum up all of the concepts and terms discussed in Chapter 8 – "New-Product Development and Product Life-Cycle Strategies". Here is your chance to make sure you know and understand the concepts!!!!

Chapter 9
Pricing: Understanding and Capturing Customer Value

Previewing the Concepts: Chapter Objectives

1. Discuss the importance of understanding customer value perceptions and company costs when setting prices.
2. Identify and define the other important internal and external factors affecting a firm's pricing decisions.
3. Describe the major strategies for pricing imitative and new products.
4. Explain how companies find a set of prices that maximize the profits from the total product mix.
5. Discuss how companies adjust their prices to take into account different types of customers and situations.
6. Discuss the key issues related to initiating and responding to price changes.

JUST THE BASICS

Chapter Overview

Pricing is the second element in the marketing mix. It plays a powerful role, and that role is detailed in this chapter. There are several sections to this chapter and a lot of material to address.

The chapter begins with discussing what a *price* actually is. It makes the point that price is more than just the money the buyer hands over to the seller—the broader view is that the price is the sum of all the values that the buyer exchanges for obtaining or using the product. There is also a brief discussion of dynamic- versus fixed-price policies, and how we as a society have evolved from dynamic to fixed and back to dynamic again.

The chapter then moves into the heart of pricing. Both internal and external factors that must be considered when setting price are detailed, as are the three general pricing approaches of cost-based pricing, value-based pricing, and competition-based pricing. The new product pricing strategies of market skimming versus market penetration are also discussed.

The chapter then moves into product mix pricing strategies. Five different strategies are outlined, including the often-forgotten category of by-product pricing. Strategies for adjusting prices, such as discount and allowance pricing, segmented pricing and psychological pricing are described, as well as initiating and responding to price changes in the marketplace. Finally, the public policy implications of pricing are covered, including the major laws that pertain to pricing.

Chapter Outline

1. **Introduction**
   a. Toys 'R' Us emerged in the late 1970s as a toy retailing "category killer," offering consumers a vast selection of toys at everyday low prices. Smaller stores and toy departments failed because they couldn't match Toys 'R' Us's selection, convenience, and low prices.
   b. In the 1990s, however, Wal-Mart offered not just everyday-low-prices on toys but rock-bottom prices.
   c. Toys 'R' Us fought back by trying to match Wal-Mart's super low prices, but with disastrous results. By early 2005 Wal-Mart held a 25 percent of the toy market; Toys 'R' Us's share had fallen to 15 percent.
   d. Toys 'R' Us now has new owners and will likely develop a new game plan. The chain is stepping back from cut-throat price wars that it can't win. It's emphasizing top-selling products and higher-margin exclusive items. It is improving store atmosphere. Still Toys 'R' Us faces an uphill battle to win back the now price-sensitive toy buyers it helped create decades ago.

2. **What Is a Price?**
   a. In the narrowest sense, price is the amount of money charged for a product or service. More broadly, price is the sum of all the values that consumers exchange for the benefits of having or using the product or service.
   b. Historically, price has been the major factor affecting buyer choice. In recent decades, nonprice factors have gained increasing importance. Price remains one of the most important elements determining a firm's market share and profitability.
   c. Price is the only element in the marketing mix that produces revenue; all other elements represent costs. Price is also one of the most flexible elements of the marketing mix.
   d. Pricing is the number one problem facing many marketing executives. But many companies do not handle pricing well. A frequent problem is that companies are too quick to reduce prices in order to get a sale rather than convincing buyers that their products are worth a higher price. Other common mistakes include pricing that is too cost-oriented and pricing that does not take the rest of the marketing mix into consideration.
   e. Smart managers treat pricing as a key strategic tools for creating and capturing customer value. Prices have a direct impact on the firm's bottom line.

3. **Factors to Consider When Setting Prices**
   a. A company's pricing decisions are affected by both internal company factors and external environmental factors, including its overall marketing strategy and mix, the nature of the market and demand, and competitors' strategies and prices.

<u>Customer Perceptions of Value</u>

b.   Pricing decisions, like other marketing mix decisions, must start with customer value. Effective customer-oriented pricing involves understanding how much value consumers place on the benefits they receive from the product and setting a price that captures this value.

c.   Value-based pricing uses buyers' perceptions of value, not the seller's cost, as the key to pricing. Value-based pricing means that the marketer cannot design a product and marketing program and then set price. Price is considered along with the other marketing mix variables before the marketing program is set.

d.   A comparison of cost-based pricing and value-based pricing is found in Figure 9-2. Cost-based pricing is product-driven. Value-based pricing reverses this process. The company sets its target price based on customer perceptions of the product value.

e.   A company using value-based pricing must find out what value buyers assign to different competitive offers. Measuring this perceived value can be difficult. Sometimes companies ask consumers how much they would pay for a basic product and for each benefit added to the offer. Or a company might conduct experiments to test the perceived value of different product offers.

f.   More and more, marketers have adopted value-pricing strategies—offering just the right combination of quality and good service at a fair price. In many cases, this has involved introducing less-expensive versions of established brand-name products.

g.   An important type of good-value pricing at the retail level is everyday low pricing (EDLP). EDLP involves charging a constant, everyday low price with few or no temporary price discounts.

   1.   In contrast, high-low pricing involves charging higher prices on an everyday basis but running frequent promotions to lower prices temporarily on selected items.

   2.   Wal-Mart practically defined this concept. To offer everyday low prices, a company must first have everyday low costs.

   3.   In many business-to-business situations, the challenge is to build the company's pricing power—its power to escape price competition and to justify higher prices and margins without losing market share. To do this, many companies adopt value-added strategies. They attach value-added services to differentiate their offers and thus support higher margins.

      i.   Often, the best strategy is not to price below the competitor, but rather to price above and convince the customers that the product is worth it.

<u>Company and Product Costs</u>

h.   Costs set the floor for the price the company can charge. The company wants to charge a price that both covers all its costs for producing, distributing, and selling the product and delivers a fair rate of return for its effort and risk.

i.   Types of costs
    1.   Fixed costs or overhead are costs that do not vary with production or sales level. Examples include rent, heat, and executive salaries.
    2.   Variable costs vary directly with the level of production. These costs tend to be the same for each unit produced.
    3.   Total costs are the sum of the fixed and variable costs for any given level of production.

Cost-based Pricing
j.   The simplest pricing method is cost-plus pricing—adding a standard markup to the cost of the product.
k.   Any pricing method that ignores demand and competitor prices is not likely to lead to the best price. Still, markup pricing remains popular for many reasons.
    1.   Sellers are more certain about costs than about demand. By tying price to cost, sellers simplify pricing.
    2.   When all firms in the industry use this pricing method, prices tend to be similar and price competition is minimized.
    3.   Many people feel that cost-plus pricing is fairer to both buyers and sellers.
l.   Break-even pricing and target-profit pricing are other cost-oriented approaches. The firm tries to determine the price at which it will break even or make the target profit it is seeking.
    1.   Target pricing uses the concept of a break-even chart, which shows the total cost and total revenue expected at different sales volume levels. See Figure 9-3 for an example. Variable costs are added to fixed cost to form total costs, which rise with each unit sold. The slope of the total revenue curve reflects the price.
    2.   However, as the price increases, demand decreases, and the market may not buy even the lower volume needed to break even at the higher price. Much depends on the relationship between price and demand. Break-even analysis and target-profit pricing do not take this relationship into account.
    3.   The company must also consider the impact of price on the sales volume needed to realize target profits and the likelihood that the needed volume will be achieved at each possible price.

Other Internal and External Factors Affecting Pricing Decisions
m.   The company must decide on its strategy for the product before setting its price. The marketing objectives include its target market and positioning; if this is set properly, then the marketing mix strategy, including price, is fairly straightforward. Pricing strategy is largely determined by decisions on market positioning.
n.   There are other general or specific objectives.

1. General objectives include survival, current profit maximization, market share leadership, or customer retention and relationship building.
2. More specifically, a company can set prices to attract new customers or to profitably retain existing ones. It can set prices low to prevent competition from entering the market or set prices at competitors' levels to stabilize the market. Prices can be set to keep the loyalty and support of resellers. Prices can be reduced temporarily to create excitement for a new product.
3. The marketing mix strategy is very important. Price decisions must be coordinated with product design, distribution, and promotion decisions to form a consistent and effective marketing program.
4. Companies often position their products on price and then tailor other marketing mix decisions to the prices they want to charge. Target costing reverses the usual process of first designing a new product, determining its cost, and then determining if they can sell it for that. Instead, it starts with an ideal selling price based on customer considerations, then targets costs that will ensure that the price is met.
5. Other companies de-emphasize price and use other marketing mix tools to create nonprice positions. Decisions about quality, promo-tion, and distribution will strongly affect price.

o. There are also organizational considerations to setting prices. Companies handle pricing in a variety of ways.
1. In small companies, prices are often set by top management.
2. In large companies, pricing is typically handled by divisional or product line managers.
3. In industrial markets, salespeople may be allowed to negotiate with customers within certain price ranges.
4. In industries where pricing is a key factor, such as aerospace and steel, companies may have a pricing department to set the best prices or help others in setting them.
5. Others who may have an influence on pricing include sales managers, production managers, finance managers, and accountants.

p. Good pricing starts with an understanding of how customers' perceptions of value affect prices. All buyers balance the price of a product or service against the benefits of owning it. Before setting prices, marketers must understand the relationship between price and demand for their products.

q. Pricing freedom varies with the different types of markets. Economists recognize four types of markets.
1. Pure competition is a market that consists of many buyers and sellers trading in a uniform commodity such as wheat, copper, or financial securities.
   i. No single buyer or seller has much effect on the going market price. Sellers in these markets do not spend much time on marketing strategy.

2. Under monopolistic competition, the market consists of many buyers and sellers who trade over a range of prices rather than a single market price.
   i. A range of prices occurs because sellers can differentiate their offers to buyers. Either the physical product can be varied in quality, features, or style, or the accompanying services can be varied.
   ii. Buyers see differences in sellers' products and will pay different prices for them.
   iii. Because there are many competitors in such markets, each firm is less affected by competitors' pricing strategies than in oligopolistic markets.
3. In oligopolistic markets, there are a few sellers who are highly sensitive to each other's pricing and marketing strategies.
   i. The product can be uniform, as in steel, or nonuniform, as in cars and computers.
   ii. There are few sellers because it is difficult for new sellers to enter the market.
4. In a pure monopoly, the market consists of one seller.
   i. The seller may be a government monopoly (U.S. Postal Service), a private regulated monopoly (a power company), or a private nonregulated monopoly (DuPont when it introduced nylon).
   ii. Pricing is handled differently in each case.
5. In a regulated monopoly, the government permits the company to set rates that will yield a "fair return," one that will let the company maintain and expand its operations as needed.
6. Nonregulated monopolies are free to set prices at what the market will bear.

r. An analysis of the price-demand relationship shows that each price the company might charge will lead to a different level of demand. The relationship between the price charged and the resulting demand level is shown in the demand curve in Figure 9-4.
   1. The demand curve shows the number of units the market will buy in a given time period at different prices that might be charged.
   2. In normal cases, demand and price are inversely related; that is, the higher the price, the lower the demand.
   3. In the case of prestige goods, the demand curve sometimes slopes upward. Consumers think that higher prices mean more quality. Still, if a company charges too high a price, the level of demand will be lower.
s. Most companies try to measure their demand curves by estimating demand at different prices. The type of market makes a difference.
   1. In a monopoly, the demand curve shows the total market demand resulting from different prices.
   2. If the company faces competition, its demand at different prices will depend on whether competitors' prices stay constant or change

with the company's own prices.

t. Price elasticity is how responsive demand will be to a change in price. If demand hardly changes with a small change in price, demand is inelastic. If demand changes greatly, demand is elastic.

  1. If demand is elastic rather than inelastic, sellers will consider lowering their price. A lower price will produce more total revenue. This practice makes sense as long as the extra costs of producing and selling more do not exceed the extra revenue.

  2. Most firms want to avoid pricing that turns their products into commodities. Marketers need to differentiate their offerings when competitors are selling the same product at a comparable or lower price.

  3. Companies need to understand the price sensitivity of their customers and the trade-offs people are willing to make between price and product characteristics.

Competitors' Strategies and Prices

u. In setting prices, the company must also consider competitors' costs, prices, and market offerings. Consumers will base their judgments of a product's value on the prices that competitors charge for similar products.

  1. The company's pricing strategy may affect the nature of the competition it faces. A high-price high margin strategy may affect competition. A low-price, low-margin strategy may stop competitors or drive them out of the market.

  2. In assessing competitors' pricing strategies, the company should ask several questions.

    i. How does the company's market offerings compare with competitors' offerings in terms of customer value?

    ii. How strong are current competitors and what are their pricing strategies?

    iii. How does the competitive landscape influence customer price sensitivity?

v. The more information customers have about competing products and prices before buying, the more price sensitive they will be. Customers will be more price sensitive if they can switch from one product to another.

w. Other factors in the company's external environment must be considered.

  1. Economic conditions can have a strong impact on the firm's pricing strategies. Boom or recession, inflation, and interest rates affect pricing because they affect both the customer perceptions of the product's price and value and the costs of producing a product.

  2. The company should set prices that give resellers a fair profit, encourage their support, and help them sell the product effectively.

  3. The government is another important external influence on prices.

  4. Short-term sales, market share, and profit goals may have to be tempered by broader societal considerations.

x.   A company sets a pricing structure that covers different items in its line. This structure changes over time as the product passes through its life cycle. As the competitive environment changes, the company considers when to initiate price changes and when to respond to them.

4. New-Product Pricing Strategies

Market-Skimming Pricing

a.   Many companies that invent new products set high prices to "skim" revenues layer by layer from the market. This is called market-skimming pricing.

b.   Market skimming makes sense only under certain conditions. First, the product's quality and image must support its higher price, and enough buyers must want the product at that price. Second, the costs of producing a smaller volume cannot be so high that they cancel the advantage of charging more. Finally, competitors should not be able to enter the market easily and undercut the high price.

Market-Penetration Pricing

c.   Market-penetration pricing sets an initial low price in order to penetrate the market quickly and deeply—to attract a large number of buyers quickly and win a large market share. The high sales volume results in falling costs, allowing the company to cut is price even further.

d.   The market must be highly price sensitive so that a low price produces more market growth. Production and distribution costs must fall as sales volume increases. The low price must help keep out competition, and the company that uses penetration pricing must maintain its low-price position.

5. Product Mix Pricing Strategies

a.   The strategy for setting a product's price often has to be changed when the product is part of a product mix. In this case, the firm looks for a set of prices that maximizes the profits on the total product mix. Pricing is difficult because the various products have related demand and costs and face different degrees of competition. See Table 9.1.

Product Line Pricing

b.   Companies usually develop product lines rather than single products. In product line pricing, management must decide on the price steps to set between the various products in a line.

c.   The price steps should take into account cost differences between the products in the line, customer evaluations of their different features, and competitors' prices.

d.   In many industries, sellers use well-established price points for the products in their line.

e.   The seller's task is to establish perceived quality differences that support the price differences.

f.   Optional-product pricing is offering to sell optional or accessory products along with the main product. For example, a car buyer may choose to order power windows, cruise control, and a CD changer.

g.   Pricing these options is a sticky problem. Using the previous example, automobile companies have to decide which items to include in the base price and which to offer as options.

Captive-Product Pricing

h.   Companies that make products that must be used along with a main product are using captive-product pricing. Producers of the main products often price them low and set high markups on the supplies.

i.   In the case of services, this strategy is called two-part pricing. The price of the service is broken into a fixed fee plus a variable usage cost. For example, theaters charge admission, then generate additional revenues from concessions. The service firm must decide how much to charge for the basic service and how much for the variable usage. The fixed amount should be low enough to induce usage of the service; profit can be made on the variable fees.

By-product Pricing

j.   In producing many commodities, such as processed meats and petroleum products, there are often by-products. Using by-product pricing, the manufacturer will seek a market for these by-products and should accept any price that covers more than the cost of storing and delivering them.

k.   By-products can even turn out to be profitable.

Product Bundle Pricing

l.   Sellers often combine several of their products and offer the bundle at a reduced price.

m.   Price bundling can promote the sales of products consumers might not otherwise buy, but the combined price must be low enough to get them to buy the bundle.

6.   **Price-Adjustment Strategies**

a.   Companies usually adjust their basic prices to account for various customer differences and changing situations. See Table 9.2.

Discount and Allowance Pricing

b.   Most companies adjust their basic price to reward customers for certain responses, such as early payment of bills, volume purchases, and off-season buying.

1.   The many forms of discounts include a cash discount, which is a price reduction to buyers who pay their bills promptly. A quantity discount is a price reduction to buyers who buy large volumes. A functional discount (also called a trade discount) is offered by the seller to trade-

channel members who perform certain functions, such as selling, storing, and record keeping. A seasonal discount is a price reduction to buyers who buy merchandise or services out of season.

2. Allowances are another type of reduction from list price. Trade-in allowances are price reductions given for turning in an old item when buying a new one. Promotional allowances are payments or price reductions to reward dealers for participating in advertising and sales support programs.

Segmented Pricing

c. In segmented pricing, the company sells a product or service at two or more prices, even though the difference in prices is not based on differences in costs.

1. Under customer-segment pricing, different customers pay different prices for the same product or service.
2. Under product-form pricing, different versions of the product are priced differently, but not according to differences in their costs.
3. Using location pricing, a company charges different prices for different locations, even though the cost of offering each location is the same.
4. Using time pricing, a firm varies its price by the season, the month, the day, and even the hour.

d. Segmented pricing can be called revenue management or yield management.

e. For segmented pricing to be effective, the market must be segmentable, and the segments must show different degrees of demand. The costs of segmenting and watching the market cannot exceed the extra revenue from the price difference. The segmented price must be legal. And segmented prices should reflect real differences in customers' perceived value.

Psychological Pricing

f. In using psychological pricing, sellers consider the psychology of prices and not simply the economics. Consumers usually perceive higher-priced products as having higher quality; when they cannot judge quality because they lack the information or skill, price becomes an important quality signal.

g. Reference prices are prices that buyers carry in their minds and refer to when looking at a given product. The reference price might be formed by noting current prices, remembering past prices, or assessing the buying situation. For most purchases, consumers don't have the skill or information to figure out whether they are paying a good price. Pricing cues are provided by the sellers.

<u>Promotional Pricing</u>

h.   With promotional pricing, companies will temporarily price their products below list price and sometimes even below cost to create buying excitement and urgency.

i.   Supermarkets and department stores will price a few products as loss leaders to attract customers to the store in the hope that they will buy other items at normal markups.

j.   Sellers will also use special-event pricing in certain seasons to draw more customers.

k.   Manufacturers sometimes offer cash rebates to consumers who buy a product from dealers within a specified time; the manufacturer sends the rebate directly to the customer.

l.   Some manufacturers offer low-interest financing, longer warranties, or free maintenance to reduce the customer's "price." Or the seller may simply offer discounts from normal prices to increase sales and reduce inventories.

m.   Promotional pricing can have adverse effects. Used too frequently and copied by competitors, price promotions can create "deal-prone" customers who wait until brands go on sale before buying them. Or, constantly reduced prices can erode a brand's value in the eyes of customers. The frequent use of promotional pricing can also lead to industry price wars.

<u>Geographical Pricing</u>

n.   A company also must decide how to price its products for customers located in different parts of the country.

o.   FOB-origin pricing means that goods are placed free on board a carrier. At that point the title and responsibility pass to the customer, who pays the freight from the factory to the destination.

p.   Uniform-delivered pricing is the opposite of FOB pricing. Here, the company charges the same price plus freight to all customers, regardless of their locations. The freight charge is set at the average freight cost. This is fairly easy to administer, and it lets the firm advertise its price nationally.

q.   Zone pricing falls between FOB-origin pricing and uniform-delivered pricing. The company sets up two or more zones. All customers within a given zone pay a single total price; the more distant the zone, the higher the price.

r.   Basing-point pricing is when the seller selects a given city as a "basing point" and charges all customers the freight cost from that city to the customer location, regardless of the city from which the goods are actually sent. Some companies set up multiple basing points to create more flexibility; they quote freight charges from the basing-point city nearest to the customer.

s.   Finally, the seller who is anxious to do business with a certain customer or geographical area might use freight-absorption pricing. Using this strategy, the seller absorbs all or part of the actual freight charges in order to get the desired business.

Dynamic Pricing

t. Companies are now using dynamic pricing---adjusting prices continually to meet the characteristics and needs of individual customers and situations.

u. Dynamic pricing allows Internet sellers to mine their databases to gauge specific shopper's desires, measure a customer's means, tailor products to fit the shopper's behavior, and price products accordingly.

v. Direct marketers monitor inventories, costs, and demand at any given moment and adjust prices instantly.

w. Buyers also benefit from the Web and dynamic pricing. A wealth of Web sites give instant product and price comparisons and let shoppers search by products and brands. Buyers can also negotiate at online auction sites.

International Pricing

x. Companies that market their products internationally must decide what prices to charge in the different countries in which they operate. In some case, a company can set a uniform worldwide price. Most companies adjust their prices to reflect local market conditions and cost considerations.

y. The price that a company should charge in a specific country depends on many factors, including economic conditions, competitive situations, laws and regulations, and development of the wholesaling and retailing systems. Consumer perceptions and preferences also may vary from country to country, calling for different prices. Or the company may have different marketing objectives in various world markets, which require changes in pricing strategy.

z. Costs play an important role in setting international prices. In some cases, price escalation may result from differences in selling strategies or market conditions. In most instances, it is simply a result of the higher costs of selling in another country—the additional costs of product modifications, shipping and insurance, import tariffs and taxes, exchange-rate fluctuations, and physical distribution.

1. More detail on international pricing is presented in Chapter 15.

7. **Price Changes**

a. After developing their pricing structures and strategies, companies often face situations in which they must initiate price changes or respond to price changes by competitors.

<u>Initiating Price Changes</u>

b.   Several situations may lead a firm to consider cutting its price. One reason is excess capacity. In this case, the firm needs more business and cannot get it through increased sales effort, product improvement, or other measures.

c.   Another situation leading to price changes is falling market share in the face of strong price competition. A company may also cut prices in a drive to dominate the market through lower costs. Either the company starts with lower costs than its competitors, or it cuts prices in the hope of gaining market share that will further cut costs through larger volume.

d.   A successful price increase can greatly increase profits. A major factor in price increases is cost inflation. Rising costs squeeze profit margins and lead companies to pass cost increases along to customers.

e.   Another factor leading to price increases is overdemand; when a company cannot supply all that its customers need, it can raise its prices, ration products to customers, or both.

f.   Prices can be raised almost invisibly by dropping discounts and adding higher-priced units to the line.

g.   In passing price increases on to customers, the company must avoid being perceived as a price gouger. There are some techniques for avoiding this problem. One is to maintain a sense of fairness surrounding any price increase. Price increases should be supported by company communications telling customers why prices are being increased.

h.   Making low-visibility price moves first is a good technique—eliminating discounts, increasing minimum order sizes, and curtailing production of low-margin products.

i.   A company should try to meet higher costs of demand without raising prices. It can consider more cost-effective ways to produce or distribute its products. It can shrink the product instead of raising the price. It can substitute less expensive ingredients or remove certain product features, packaging, or services. Or it can "unbundle" its products and services, removing and separately pricing elements that were formerly part of the offer.

j.   Whether the price is raised or lowered, the action will affect buyers, competitors, distributors, and suppliers, and may interest the governments well.

     1.   Customers do not always interpret prices in a straightforward way. They may view a price cut in several ways. They might believe that quality was reduced. Or they might think that the price will come down even further and that it will pay to wait and see.

     2.   A price increase may have some positive meanings for buyers. Customers might think that the item is very "hot" and may be unobtainable unless they buy it soon. Or they might think that the item is an unusually good value.

3.   Competitors are most likely to react to a price change when the number of firms involved is small, when the product is uniform, and when the buyers are well informed.

4.   Like with a consumer, a competitor can interpret price changes in many ways. So the company must guess each competitor's likely reaction.

Responding to Price Changes

k.   In responding to competitors' price changes, the company needs to consider several issues: Why did the competitor change the price? Was it to take more market share, to sue excess capacity, to meet changing cost conditions, or to lead an industry-wide price change? Is the price change temporary or permanent? What will happen to the company's market share and profits if it does not respond? Are other companies going to respond?

l.   Planning ahead for price changes cuts down reaction time. Figure 9-5 shows the ways a company might assess and respond to a competitor's price cut.

1.   It could reduce its price to match the competitor's price. It may decide the market is price sensitive and that it would lose too much market share to the lower-price competitor.

2.   The company could maintain its price but raise the perceived value of its offer. It could improve communications, stressing the relative quality of its product over that of the lower-price competitor.

3.   The company might improve quality and increase price, moving its brand into a higher-price position.

4.   The company might launch a low-price "fighting brand" by adding a lower-price item to the line or creating a separate lower-price brand. This is necessary if the particular market segment being lost is price sensitive and will not respond to arguments of higher quality.

8.   **Public Policy and Pricing**

a. Price competition is a core element of our free-market economy. In setting prices, companies are not usually free to charge whatever prices they wish. Many federal, state, and even local laws govern the rules of fair play in pricing. The most important pieces of legislation affecting pricing are the Sherman, Clayton, and Robinson-Patman acts, initially adopted to curb the formation of monopolies and to regulate business practices that might unfairly restrain trade.

b.   Figure 9-6 shows the major public policy issues in pricing. These include price-fixing and predatory pricing as well as retail price maintenance, discriminatory pricing, and deceptive pricing.

Pricing Within Channel Levels

c.   Federal legislation on price-fixing states that sellers must set prices without talking to competitors. Otherwise, price collusion is suspected.

d.   Sellers are also prohibited from using predatory pricing—selling below cost with the intention of punishing a competitor or gaining higher long-run profits by putting competitors out of business. This protects small sellers from larger ones who might sell items below cost temporarily or in a specific locale to drive them out of business.

Pricing Across Channel Levels

e.   The Robinson-Patman Act seeks to prevent unfair price discrimination by ensuring that sellers offer the same price terms to customers at a given level of trade. Every retailer is entitled to the same price terms from a given manufacturer. However, price discrimination is allowed if the seller can prove that its costs are different when selling to different retailers. Or the seller can discriminate in its pricing if the seller manufactures different qualities of the same product for different retailers.

f.   Retail price maintenance is also prohibited—a manufacturer cannot require dealers to charge a specified retail price for its products. Although the seller can propose a manufacturer's suggested retail price to dealers, it cannot refuse to sell to a dealer who takes independent pricing action, nor can it punish a dealer by shipping late or denying advertising allowances.

g.   Deceptive pricing occurs when a seller states prices or price savings that mislead consumers or are not actually available to consumers. This might involve bogus reference or comparison prices, as when a retailer sets artificially high "regular" prices then announces "sale" prices close to its previous everyday prices.

  1.   Scanner fraud is another means of deceptive pricing. The widespread use of scanner-based computer checkouts has led to increasing complaints of retailers overcharging their customers.

  2.   Price confusion results when firms employ pricing methods that make it difficult for consumers to understand just what price they are really paying. For example, consumers are sometimes misled regarding the real price of a home mortgage or car leasing agreement.

Creative Marketing Exercises Designed to Reinforce the Concepts!!! (Suggested answers to these exercises can be found at the end of the Study Guide.)

1. Visit www.sheraton.com and search for a room in Miami, Florida during the month of June. How do you account for the variety of room rates?
2. Identify 5 products that you think are priced too low. Rationalize your answers.
3. Identify 5 products you think are priced too high. Rationalize your answer.
4. How do non-profit organizations price their products? If profit is not the basis, then what is?
5. What enticed you to buy your last impulse item? Outline the justification for your purchase.
6. Conduct a brief research study on Mexican labor laws. Why are so many of our manufacturing companies relocating to Mexico?
7. List 5 products that are priced high at this moment in time but the price will drop over the next 5 years.
8. Create a poster with 5 sections with pictures to represent each of the following: pure competition, monopolistic competition, oligopolistic competition, and pure monopoly.
9. Create a list of items where you think demand is elastic and another list where you think demand is inelastic. Draw a comparison between the two lists.
10. Go to www.eddiebauer.com and conduct a search of their sale items. Outline how this strategy is an example of high-low pricing.

"Linking the Concepts" (#1) -- Suggestions/Hints

1. Most consumers would not see this product as offering great value. The Bentley Continental GT is a more expensive car than the average consumer would be willing to pay. Even though the automobile had some nice accessories, it is still out of most consumers' price range for an auto. It is still perceived as a luxury vehicle. (It is important to note, however, that there is a lengthy waiting list for those consumers wishing to purchase this vehicle).

2. In the restaurant industry, there are some extremely expensive places to eat and some inexpensive places. Compare one of the restaurants with a famous chef and a McDonald's. The expensive restaurant sells image, the experience, and the perception of the meal. There may not even be a large quantity of food on the plate, but it is presented well. Conversely, McDonald's gives more food per dollar, but the presentation is not as exclusive.

   For the expensive restaurant, it offers the best value when it comes to presentation. For McDonald's, it offers the best value when it comes to food quantity.

3. No, value does not mean the same as "low price." Value is giving the consumer more that what is expected for the price paid. There is the example of a fast and expensive sports car; it may give the consumer greater value if it goes faster than they had hoped. For a grocery store, they might offer consumers value because the customer can purchase more groceries for their money than at a competitor's store. The customer determines how value will be defined.

"Linking the Concepts" (#2) -- Suggestions/Hints

1. In the fast food industry, for example, companies have tried to slow the "deal prone" onslaught by creating value meals and value menu items at set prices. These products are set at a low value-added price and they are not discounted anymore. This way consumers can get acquainted with a set lower price and know what to expect upon their purchase. The consumers know the price and are not looking for deals on those items. They are already pleased with the set value prices.

2. "Value" has become a promotional tool because it always seems to be tied to some promotion. It has trained some consumers into thinking that value is only "low prices." This is not how "value" is defined, however. Promotional pricing adds value for the consumer who is looking to get a lot of "bang for their buck." However, if the consumer is image and quality based, then promotional pricing may take away from those perspectives. For example, a Porsche purchaser is not as moved by promotional pricing as he/she might be just for the image of owning the car.

Marketing Adventure Exercises (Suggested answers to these exercises can be found at the end of the Study Guide.)

(Visit www.prenhall.com/adventure for advertisements.)

1. Electronics   Sony

Consider the product in the selected ad. Is it positioned on a price or non-price basis? Explain your answer.

2. Apparel   Student choice

Describe the marketplace under monopolistic competition. Choose a product in the Apparel section that you believe represents this type of marketplace.

3.   Travel   Student choice

What are the characteristics of an oligopoly? Which of the products in the Travel category would compete in this type of market?

4.   Auto   Student choice

What is the difference between cost-based pricing and value-based pricing? Select an advertisement from the Auto category and explain its type of pricing.

5.   Student choice

Discuss the differences between market-skimming and market-penetration pricing. Select an advertisement for each strategy and explain why the product would use this pricing strategy.

6.   Auto   Cadillac
      Electronics   Olympus, Samsung
      Exhibits   Zurich Opera
      Food   Perdue

There are five product mix pricing situations. List and describe each. Match each to the most appropriate product in the selected ads.

7.   Apparel   Timberland

List and explain the different types of discount pricing. Explain how each might be used in relation to the selected ad.

8.   Electronics   Samsung
      Exhibits   Toronto
      Services   U_of_Natal
      Travel   Quantas

What is segmented pricing? Match the selected ads with the type of segmented pricing that is most appropriate for each.

SUM IT UP!!!!!!

Using only this page, sum up all of the concepts and terms discussed in Chapter 9 – "Pricing: Understanding and Capturing Customer Value." Here is your chance to make sure you know and understand the concepts!!!!

Chapter 10
Marketing Channels and Supply Chain Management

Previewing the Concepts: Chapter Objectives

1. Explain why companies use distribution channels and discuss the functions these channels perform.
2. Discuss how channel members interact and how they organize to perform the work of the channel.
3. Identify the major channel alternatives open to a company.
4. Explain how companies select, motivate, and evaluate channel members.
5. Discuss the nature and importance of marketing logistics and integrated supply chain management.

JUST THE BASICS

Chapter Overview

This chapter covers the important topics of supply chain management. Supply chains consist of both upstream and downstream partners, including suppliers, intermediaries, and even intermediary customers. The term *value delivery network* expands on the limited nature of "supply chain." It consists of the company, suppliers, distributors, and ultimately customers who "partner" with each other to improve the performance of the entire system.

The chapter focuses on marketing channels—the downstream side of the value delivery network. A company's channel decisions directly affect every other marketing decision. And because distribution channel decisions often involve long-term commitments to other firms, management must define its channels carefully, with an eye on tomorrow's likely selling environment as well as today's.

Channel members add value by bridging the major time, place, and possession gaps that separate goods and services from those who would use them. Members of the marketing channel perform many key functions, such as gathering and distributing marketing information; promoting products; contacting prospective buyers; matching supply with demand; negotiating final prices; and performing the physical distribution of the goods, financing large purchases, and taking the risk of selling the product.

For channels to work properly, each channel member's role must be specified and conflict must be managed. Conventional distribution systems typically lacked a strong leader; vertical marketing systems (VMS) have evolved to provide that channel leadership. The three major types of VMSs include corporate, contractual, and administered.

In designing marketing channels, managers must analyze customer needs, set channel objectives, identify major channel alternatives, and then evaluate those alternatives. In designing international channels, marketers will face additional complexities. Each country has its own unique distribution system that has evolved over time and changes very slowly.

Marketing logistics, also called physical distribution, involves planning, implementing and controlling the physical flow of goods, services, and other related information from points of origin to points of consumption. It involves getting the right product to the right customer in the right place at the right time. Marketing logistics addresses not only outbound distribution, but also inbound and reverse distribution. The major logistics functions include warehousing, inventory management, and transportation.

Chapter Outline

1. **Introduction**
    a. Caterpillar believes its dominance over seven decades in the market for heavy construction and mining equipment results from its unparalleled distribution and customer support system.
    b. Caterpillar sells more than 300 products in nearly 200 countries, generating sales of more than $30 billion annually. It has 30% of the worldwide construction-equipment business, more than double that of number two Komatsu.
    c. Competitors often bypass their dealers and sell directly to big customers to cut costs or make more profits for themselves, but Caterpillar wouldn't think of going around its dealers. Caterpillar's superb distribution system serves as a major source of competitive advantage. The system is built on a firm base of mutual trust and shared dreams.
    d. Most firms cannot bring value to customers by themselves. Instead, they must work closely with other firms in a larger value delivery network.

2. **Supply Chains and the Value Delivery Network**
    a. The supply chain consists of "upstream" and "downstream" partners, including suppliers, intermediaries, and even intermediary customers.
        1. Upstream from the manufacturer or service provider is the set of firms that supply the raw materials, components, parts, information, finances, and expertise needed to create a product or service.
        2. Marketers have traditionally focused on the downstream side of the supply chain, which are the marketing channels or distribution channels that look forward toward the customer.
    b. It is the unique design of each company's supply chain that enables it to deliver superior value to customers.
        1. The term *supply chain* may be too limited—it takes a make-and-sell view of the business.

2. A better term would be *demand chain* because it suggests a sense-and-respond view of the market. Under this view, planning starts with the needs of the target customers, to which the company responds by organizing resources with the goal of creating customer value.

3. Even this might be too limiting, however. A value delivery network is made up of the company, suppliers, distributors, and ultimately customers who partner with each other to improve the performance of the entire system.

c. This chapter focuses on marketing channels—on the downstream side of the value delivery network.

d. There are four major questions concerning marketing channels:
1. What is the nature of marketing channels and why are they important?
2. How do channel firms interact and organize to do the work of the channel?
3. What problems do companies face in designing and managing their channels?
4. What role do physical distribution and supply chain management play in attracting and satisfying customers?

3. **The Nature and Importance of Marketing Channels**

a. A marketing channel or distribution channel is a set of interdependent organizations involved in the process of making a product or service available for use or consumption by the consumer or business user.

b. A company's channel decisions directly affect every other marketing decision.

c. Distribution channel decisions often involve long-term commitments to other firms. Therefore, management must design its channels carefully, with an eye on tomorrow's likely selling environment as well as today's.

How Channel Members Add Value

d. The use of intermediaries results from their greater efficiency in making goods available in target markets. Through their contacts, experience, specialization, and scale of operation, intermediaries usually offer the firm more than it can achieve on its own.

e. Figure 10-1 shows how using intermediaries can provide economies.

f. The role of marketing intermediaries is to transform the assortments of products made by producers into the assortments wanted by consumers.
1. Producers make narrow assortments of products in large quantities, but consumers want broad assortments of products in small quantities.
2. Intermediaries play an important role in matching supply and demand.

g.   Channel members add value by bridging the major time, place, and possession gaps that separate goods and services from those who would use them.

h.   Members of the marketing channel perform many key functions:
       1.   Information: gathering and distributing marketing research and intelligence.
       2.   Promotion: developing and spreading persuasive communications about an offer.
       3.   Contact: finding and communicating with prospective buyers.
       4.   Matching: shaping and fitting the offer to the buyer's needs.
       5.   Negotiation: reach an agreement on price and other terms of the offer.
       6.   Physical distribution: transporting and storing goods.
       7.   Financing: acquiring and using funds to cover the costs of the channel work.
       8.   Risk taking: assuming the risks of carrying out the channel work.

i.   In dividing the work of the channel, the various functions should be assigned to the channel members who can add the most value for the cost.

Number of Channel Levels

j.   Each layer of marketing intermediaries that performs some work in bringing the product and its ownership closer to the final buyer is a channel level.

k.   The number of intermediary levels indicates the length of a channel. Figure 10-2A shows several distribution channels of different lengths.
       1.   A direct marketing channel has no intermediary levels; the company sells directly to consumers.
       2.   An indirect marketing channel contains one or more intermediaries.

l.   Figure 10-2B shows some common business distribution channels.
       1.   A business marketer can use its own sales force to sell directly to business customers.
       2.   Or it can sell to various types of intermediaries, who in turn sell to these customers.

m.   From the producer's point of view, a greater number of levels means less control and greater channel complexity.

n.   The institutions in the channel are connected by several types of flows.
       1.   The flows include physical flow of the products, the ownership flow, the payment flow, the information flow, and the promotion flow.
       2.   These flows can make even channels with only one or a few levels very complex.

4. **Channel Behavior and Organization**
    a. Distribution channels are complex behavioral systems in which people and companies interact to accomplish individual, company, and channel goals.
        1. Some channel systems consist only of informal interactions among loosely organized firms.
        2. Others consist of formal interactions guided by strong organizational structures.

Channel Behavior
    b. Each channel member plays a specialized role in the channel. The channel will be most effective when each member is assigned the tasks it can do best.
    c. Ideally, all channel firms should work together smoothly. They should understand and accept their roles, coordinate their activities, and cooperate to attain overall channel goals.
    d. Although channel members depend on one another, they often act alone in their own short-run best interests. Disagreements over goals, roles, and rewards generate channel conflict.
        1. Horizontal conflict occurs among firms at the same level of the channel.
        2. Vertical conflict is more common; it is conflict between different levels of the same channel.
        3. Some conflict in the channel takes the form of healthy competition.
        4. Severe or prolonged conflict can disrupt channel effectiveness and cause lasting harm to channel relationships.

Vertical Marketing Systems
    e. Historically, conventional distribution systems have lacked leadership and power, often resulting in damaging conflict and poor performance.
    f. One of the biggest channel developments over the years has been the emergence of vertical marketing systems that provide channel leadership. Figure 10-3 contrasts the two types of channel arrangements.
        1. A conventional distribution channel consists of one or more independent producers, wholesalers, and retailers. Each is a separate business seeking to maximize its own profits.
        2. A vertical marketing system (VMS) consists of producers, wholesalers, and retailers acting as a unified system. One channel member owns the others, has contracts with them, or wields so much power that they must all cooperate. The VMS can be dominated by either the producer, the wholesaler, or the retailer.
            i.   A corporate VMS integrates successive stages of production and distribution under single ownership.
            ii. A contractual VMS consists of independent firms at different levels of production and distribution who join together through contracts to obtain more economies or sales impact than each could achieve alone.

a. The franchise organization is the most common type. There are three types of franchises: manufacturer-sponsored retailer franchiser system; manufacturer-sponsored wholesaler franchise system; and a service-firm-sponsored retailer franchiser system.

    iii. An administered VMS is one where leadership is assumed not through common ownership or contractual ties but through the size and power of one or a few dominant channel members.

Horizontal Marketing Systems

g. A horizontal marketing system is one in which two or more companies at one level join together to follow a new marketing opportunity. By working together, companies can combine their financial, production, or marketing resources to accomplish more than any one company could alone.

h. Companies might join forces with competitors or noncompetitors. They might work with each other on a temporary or permanent basis, or they may create a separate company.

Multichannel Distribution Systems

i. More and more companies have adopted multichannel distribution systems, which are also called hybrid marketing channels.

j. This occurs when a single firm sets up two or more marketing channels to reach one or more customer segments.

k. Figure 10-4 shows a hybrid channel. These days almost every large company and many small ones distribute through multiple channels.

l. With each new channel, the company expands its sales and market coverage and gains opportunities to tailor its products and services to the specific needs of diverse customer segments.

m. Multichannel systems are harder to control and they generate conflict as more channels compete for customers and sales.

Changing Channel Organizations

n. One major trend is toward disintermediation—more and more, product and service producers are bypassing intermediaries and going directly to final buyers, or to radically new types of channel intermediaries.

o. This presents problems and opportunities.
    1. To avoid being swept aside, traditional intermediaries must find new ways to add value in the supply chain.
    2. To remain competitive, product and service producers must develop new channel opportunities, such as Internet and other direct channels. Developing these channels brings them into direct competition with their established channels, resulting in conflict.

5. **Channel Design Decisions**

   a.   In designing marketing channels, manufacturers struggle between what is ideal and what is practical.

   1.   A new firm with limited capital usually starts by selling in a limited market area.

   2.   In this way, channel systems often evolve to meet market opportunities and conditions.

   b.   Channel analysis and design should be purposeful—decision making should include analyzing consumer needs, setting channel objectives, identifying major channel alternatives, and evaluating those alternatives.

Analyzing Customer Needs

   c.   Marketing channels are part of the overall customer value delivery network. Thus, designing the marketing channel starts with finding out what target consumers want from the channel.

   d.   The company must balance consumer needs not only against the feasibility and costs of meeting these needs, but also against customer price preferences.

Setting Channel Objectives

   e.   Companies should state their marketing channel objectives in terms of targeted levels of customer service. In each segment, the company wants to minimize the total channel cost of meeting customer service requirements.

   f.   The company's channel objectives are influenced by the nature of the company, its products, its marketing intermediaries, its competitors, and the environment.

   g.   Environmental factors such as economic conditions and legal constraints may affect channel objectives and design.

Identifying Major Alternatives

   h.   The company should next identify its major channel alternatives in terms of types of intermediaries, the number of intermediaries, and the responsibilities of each channel member.

   1.   A firm should identify the types of channel members available to carry out its channel work.

   i.   Company sales force

   ii. Manufacturer's agency

   iii. Industrial distributors

   2.   Companies must also determine the number of channel members to use at each level.

   i.   Intensive distribution is a strategy in which they stock their products in as many outlets as possible.

   ii. In exclusive distribution, the producer gives only a limited number of dealers the exclusive right to distribute its product in their territories.

        iii.   In between intensive and exclusive distribution is selective distribution—the use of more than one, but fewer than all, of the intermediaries who are willing to carry a company's products.

    3.   The producer and intermediaries need to agree on the terms and responsibilities of each channel member.

        i.   They should agree on price policies, conditions of sale, territorial rights, and specific services to be performed by each party.

        ii.   Mutual services and duties need to be spelled out carefully.

Evaluating the Major Alternatives

i.   Each alternative should be evaluated against economic, control, and adaptive criteria.

    1.   Using economic criteria, a company compares the likely sales, costs, and profitability of different channel alternatives.

    2.   The company must also consider control issues. Using intermediaries usually means giving them some control over the marketing of the product, and some intermediaries take more control than others.

    3.   The company must also apply adaptive criteria. Channels often involve long-term commitments, yet the company wants to keep the channel flexible so that it can adapt to environmental changes.

Designing International Distribution Channels

j.   International marketers face many additional complexities in designing their channels.

k.   Each country has its own unique distribution system that has evolved over time and changes very slowly.

    1.   In some markets, the distribution system is complex and hard to penetrate, consisting of many layers and large numbers of intermediaries (e.g., Japan).

    2.   At the other extreme, distribution systems in developing countries may be scattered and inefficient, or altogether lacking (e.g., China and India).

6.   **Channel Management Decisions**

a.   Once the company has reviewed its channel alternatives and decided on the best channel design, it must implement and managed the chosen channel.

b.   Channel management calls for selecting, managing, and motivating individual channel members and evaluating their performance over time.

Selecting Channel Members

c.   Producers vary in their ability to attract qualified marketing intermediaries.

d. When selecting intermediaries, the company should determine what characteristics distinguish the better ones. It will want to evaluate each channel member's years in business, other lines carried, growth and profit record, cooperativeness, and reputation.

Managing and Motivating Channel Members

e. Once selected, channel members must be continuously managed and motivated to do their best.
    1. The company must sell not only through the intermediaries but to and with them.
    2. They practice strong partner relationship management (PRM) to forge long-term partnerships with channel members.

f. In managing its channels, a company must convince distributors that they can succeed better by working together as a part of a cohesive value delivery system.
    1. Many companies are now installing integrated high-tech partner relationship management systems to coordinate their whole-channel marketing efforts.
    2. Companies now use PRM and supply chain management (SCM) software to help recruit, organize, manage, motivate, and evaluate relationships with channel partners.

Evaluating Channel Members

g. The producer must regularly check channel member performance against standards such as sales quotas, average inventory levels, customer delivery time, treatment of damaged and lost goods, cooperation in company promotion and training programs, and services to the customer.

h. The company should recognize and reward intermediaries who are performing well and adding good value for consumers. Those who are performing poorly should be assisted or replaced.

7. **Public Policy and Distribution Decisions**
    a. For the most part, companies are legally free to develop whatever channel arrangements suit them.
    b. Laws affecting channels seek to prevent the exclusionary tactics of some companies that might keep another company from using a desired channel.

Exclusive Dealing

c. Exclusive distribution occurs when the seller allows only certain outlets to carry its products. When the seller requires that these dealers not handle competitors' products, its strategy is called exclusive dealing.
    1. Both parties can benefit from exclusive arrangements.
    2. But exclusive arrangements also exclude other producers from selling to these dealers.
        i. This brings exclusive dealing contracts under the scope of the Clayton Act of 1914.

          ii.   They are legal as long as they do not substantially lessen competition or tend to create a monopoly, and as long as both parties enter into the agreement voluntarily.

     3.   Exclusive dealing often includes exclusive territorial agreements.

d.   Producers of a strong brand sometimes sell it to dealers only if the dealers will take some or all of the rest of the line. This is called full-line forcing.

     1.   These tying arrangements may not be illegal, but if they lessen competition substantially, they do come under the Clayton Act.

e.   Producers are free to select their dealers, but their right to terminate dealers is somewhat restricted.

     1.   Sellers can drop dealers "for cause."

     2.   They cannot drop dealers if, for example, the dealers refuse to cooperate in a doubtful legal arrangement.

8.   **Marketing Logistics and Supply Chain Management**

a.   Companies must decide on the best way to store, handle, and move their products and services so that they are available to customers in the right assortments, at the right time, and in the right place.

b.   Physical distribution and logistics effectiveness has a major impact on both customer satisfaction and company costs.

Nature and Importance of Marketing Logistics

c.   Marketing logistics, also called physical distribution, involves planning, implementing, and controlling the physical flow of goods, services, and related information from points of origin to points of consumption to meet customer requirements at a profit. It involves getting the right product to the right customer in the right place at the right time.

d.   Marketing logistics addresses not only outbound distribution (moving products from the factory to resellers and ultimately to customers) but also inbound distribution (moving products and materials from suppliers to the factory) and reverse distribution (moving broken, unwanted, or excess products returned by customers or resellers).

e.   It involves entire supply chain management—managing upstream and downstream value-added flows of materials, final goods, and related information among suppliers, the company, resellers, and final consumers. See Figure 10-5.

f.   The logistics manager's task is to coordinate activities of suppliers, purchasing agents, marketers, channel members, and customers.

g.   Companies can gain a powerful competitive advantage by pursuing improved logistics to give customers better service or lower prices.

     1.   Improved logistics can yield tremendous cost savings to both the company and its customers. As much as 20% of an average product's price is accounted for by shipping and transportation alone.

     2.   The explosion in product variety has created a need for improved logistics management.

3.   Improvements in information technology have created opportunities for major gains in distribution efficiency.

Goals of the Logistics System

h.   No logistics system can both maximize customer service and minimize distribution costs.
      1.   Maximum customer service implies rapid delivery, large inventories, flexible assortments, liberal return policies, and other services—all of which raise distribution costs.
      2.   Minimum distribution costs imply slower delivery, smaller inventories, and larger shipping lots, which represent a lower level of overall customer service.

i.   The goal of marketing logistics should be to provide a targeted level of customer service at the least cost.
      1.   A company must first research the importance of various distribution services to customers and then set desired service levels for each segment.
      2.   The objective is to maximize profits, not sales.

Major Logistics Functions

j.   The major logistics functions include warehousing, inventory management, transportation, and logistics information management.

k.   A company must decide on how many and what types of warehouses it needs and where they will be located.
      1.   A storage warehouse stores goods for moderate to long periods.
      2.   Distribution centers are designed to move goods rather than to store them. They are large and highly automated warehouses designed to receive goods from various plants and suppliers, take orders, fill those orders efficiently, and deliver goods to customers as quickly as possible.
      3.   New, single-storied automated warehouses have advanced, computer-controlled materials-handling systems requiring few employees. Computers and scanners read orders and direct lift trucks, electric hoists, or robots to gather goods, move them to loading docks, and issue invoices.

l.   Inventory management also affects customer satisfaction. Here, managers must maintain the delicate balance between carrying too little inventory and carrying too much.
      1.   Just-in-time logistics systems carry only small inventories of parts or merchandise, often for only a few days of operation. New stock arrives exactly when needed, rather than being stored in inventory until being used.

m.   The choice of transportation carriers affects the pricing of products, delivery performance, and condition of the goods when they arrive.
      1.   Trucks have increased their share of transportation steadily and now account for 32% of total cargo ton-miles (more than 58% of

actual tonnage). They account for the largest portion of transportation within cities as opposed to between cities.

    2.   Railroads account for 28% of total cargo ton-miles moved. They are one of the most cost-effective modes for shipping large amounts of bulk products.

    3.   Water carriers, which account for about 16% of cargo ton-miles, transport large amounts of goods by ships and barges on U.S. coastal and inland waterways. Although the cost of water transportation is very low for shipping bulky, low-value, nonperishable products, water transportation is the slowest mode and may be affected by the weather.

    4.   Pipelines are a specialized means of shipping petroleum, natural gas, and chemicals from sources to markets.

    5.   Although air carriers transport less than 1% of the nation's goods, they are an important transportation mode. Airfreight rates are much higher than rail or truck rates, but airfreight is ideal when speed is needed or distant markets have to be reached.

    6.   The Internet carries digital products from producer to customer via satellite, cable modem, or telephone wire.

    7.   Intermodal transportation is combining two or more modes of transportation.

        i.   Piggyback describes the use of rail and trucks.

        ii.   Fishyback is combining water and trucks.

        iii.   Trainship combines water and rail.

        iv.   Airtruck combines air and trucks.

n.   Companies manage their supply chains through information. Channel partners often link up to share information and to make better joint decisions.

    1.   Information can be shared and managed by mail or telephone, through salespeople, or through traditional or Internet-based electronic data interchange (EDI), the computerized exchange of data between organizations.

    3.   Suppliers might be asked to generate orders and arrange deliveries for their customers. Many retailers set up a vendor-managed inventory systems (VMI) or continuous inventory replenishment systems. Such systems require close cooperation between buyer and seller.

Integrated Logistics Management

o.   Integrated logistics management recognizes that providing better customer service and trimming distribution costs require teamwork, both inside the company and among all the marketing channel organizations.

p.   The goal of integrated supply chain management is to harmonize all of the company's logistics decisions.

1. Some companies have created permanent logistics committees made up of managers responsible for different physical distribution activities.

2. Companies can also create management positions that link the logistics activities of functional areas.

3. Companies can employ sophisticated, systemwide supply chain management software.

q. The members of a distribution channel are linked closely in delivering customer satisfaction and value as well as building customer relationships.

1. Smart companies coordinate their logistics strategies and forge strong partnerships with suppliers and customers to improve customer service and reduce channel costs.

2. Many companies have created cross-functional, cross-company teams.

3. Other companies partner through shared projects.

r. Integrated logistics companies, called third-party logistics (3PL) providers, perform any or all of the functions required to get their clients' product to market.

1. Companies use third-party logistics providers for many reasons.

i. Getting the product to market is the focus of the logistics provider, so these providers can often do it more efficiently and at lower cost.

ii. Outsourcing logistics frees a company to focus more intensely on its core business.

iii. Integrated logistics companies understand increasingly complex logistics environments. This can be helpful to companies attempting to expand their global market coverage.

Creative Marketing Exercises Designed to Reinforce the Concepts!!! (Suggested answers to these exercises can be found at the end of the Study Guide.)

1. Interview the manager of your favorite fast food restaurant. Outline the manner in which they receive products and ask them to determine the pros and cons of the method.

2. You are trying to decide whether or not to custom order a car from overseas. What factors would you consider?

3. Compare the price of an item online at www.qvc.com to the regular retail price. How can you explain the difference?

4. Could there be a conflict between the prices of items in a catalog and the price of the item in the store? Explain.

5. How do rental car agencies deliver their cars to the customer? Conduct a brief research study and outline their system.

6. Visit www.dol.gov and determine which job markets are deemed the most secure in the nation. Give a brief justification for the top 3.

7. Visit www.verabradley.com and discuss why this company is a good example of an exclusive distribution network.

8. Outline the U.S. Trade Commission's procedure for someone to follow if they wanted to trade with the U.S.
9. Investigate and identify 5 companies that offer international carrier services. How many ways can most companies transport goods?
10. How do restaurants keep their inventory at a functional level? Interview a restaurateur in your area and find out.

"Linking the Concepts" (#1) -- Suggestions/Hints

1. Caterpillar uses dealers to sell their products. The dealers set up relationships with Caterpillar and their own customers. There are no other levels in this distribution process. Goodyear sells through independent-dealer channels and also through mass-merchant retailers (such as Sears).

   Below are graphics showing Caterpillar and Goodyear's distribution processes:

   Caterpillar -----------dealers-------------consumers

   Goodyear -----------independent dealers---------consumers

   Goodyear-------------------mass-merchant retailers-----------consumers

2. In comparing the two companies, differences in dealer relations are quite evident. Caterpillar has made, and relies on, strong relationships with their dealers. They offer extraordinary support to the dealers, good and open communication, and create personal relationships, as well as business relationships.

   Goodyear created hard feelings and conflict with its premier independent-dealer channel when it began selling through mass-merchant retailers. As a result, Goodyear's relations with its dealers have steadily deteriorated. Also, Goodyear began to offer bulk discounts to its biggest retailers and wholesalers completely disregarding the smaller dealers. This created channel conflict, thereby disrupting channel effectiveness and cause lasting harm to channel relationships. Thus, Goodyear should manage this channel conflict to keep it from getting out of hand.

"Linking the Concepts" (#2) -- Suggestions/Hints

1. Again, Caterpillar uses dealers to sell their products. The dealers set up relationships with Caterpillar and their own customers. There are no other levels in this distribution process. Caterpillar has worked hard to establish these so-called, "hand-shake" relationships. This works for their type of products and customers.

   GE has gone to an online accessible website for their dealers. This allows 24 hours a day, 7 days a week access for the dealers, as compared to the relationships that Caterpillar has established. Dealers can check on product availability, prices, place orders, and receive order status information. They can also create custom brochures and order POP material. GE now offers next-day delivery, so dealers do not need to carry as much inventory. This creates a great relationship between the dealers and the company.

2. Both companies have been very successful at managing and supporting these channels. They rely heavily on their relationships with their dealers. Both companies stay in contact with their dealers, electronically and personally. They need to be able to respond to their dealers' needs quickly. Each company must be successful at managing their relationships and responding to customers' needs because both companies have stayed in business for a long time and dealers are still applying to be part of each company's distribution process.

Marketing Adventure Exercises (Suggested answers to these exercises can be found at the end of the Study Guide.)

(Visit www.prenhall.com/adventure for advertisements.)

1.   Apparel   Levis

Review the product in the selected ad. Discuss the company's partners in its supply chain.

2.   Student choice

The text states, "Companies often pay too little attention to their distribution channels, sometimes with damaging results." Select ads for companies that you believe have established distribution systems that provide the company with a competitive advantage.

3.   Food   Snickers, Winterfresh
        Household   Bic

Why would marketing intermediaries be needed for the products in the selected ads?

4.   Student choice

Describe the difference between a direct marketing channel and an indirect marketing channel. For each of these consumer distribution channels, select a representative ad.

5. Financial   Hrblock
    Travel   Imperial

Describe the potential channel conflicts that may exist for the product in the selected ad.

6. Autos   Ford
    Food   Pepsi
    Food   McDonalds

The text describes different types of franchises. Review those descriptions and match the selected ads to the appropriate franchise type.

7. Electronics   Motorola

Explain what a multichannel distribution system is, and its advantages and disadvantages. Relate this to the selected ad.

8. Student choice

What is disintermediation? Select ads to show examples of this and provide reasons why this phenomenon exists today.

9. Auto   Mercedes
    Cosmetics   Colgate
    Electronics   Sony

Discuss the differences between intensive distribution, selective distribution, and exclusive distribution. Match the selected ads with the appropriate distribution strategy.

SUM IT UP!!!!!!

Using only this page, sum up all of the concepts and terms discussed in Chapter 10 – "Marketing Channels and Supply Chain Management". Here is your chance to make sure you know and understand the concepts!!!!

Chapter 11
Retailing and Wholesaling

Previewing the Concepts: Chapter Objectives

1. Explain the role of retailers and wholesalers in the distribution channel.
2. Describe the major types of retailers and give examples of each.
3. Identify the major types of wholesalers and give examples of each.
4. Explain the marketing decisions facing retailers and wholesalers.

JUST THE BASICS

Chapter Overview

This chapter is a continuation of the prior chapter on marketing channels; it provides more detail on retailing and wholesaling, two very important concepts in the value delivery network.

It begins with a discussion of retailers and the challenges they face. There are many types of retailers. These retailers can be classified according to several characteristics, including the amount of service they offer, the breadth and depth of their product lines, the relative prices they charge, and how they are organized.

Retailers are always searching for new strategies to attract and retain customers. The major decisions retailers need to make are centered around their target market and positioning, their product assortment and services, their price, their promotion strategies, and where they are located.

Retailing is facing many challenges, including new retail forms, such as warehouse stores. The wheel of retailing concept says that many new retailing forms begin as low-margin, low-price, low-status operations. They challenge established retailers, and then the new retailers' success leads them to upgrade their facilities and offer more service. In turn, their costs increase, and eventually they become like the conventional retailers they replaced. The cycle begins again.

Wholesalers buy mostly from producers and sell mostly to retailers, industrial customers, and other wholesalers. As a result, many of this country's largest and most important wholesalers are largely unknown to final consumers. Wholesalers provide important services, however, and they add value through performing one or more of several functions.

There are many types of wholesalers, including merchant wholesalers, agents and brokers, and manufacturers' sales branches and offices. They face many of the same decisions as retailers, including the choice of target market, positioning, and the marketing mix.

The distinction between large retailers and large wholesalers continues to blur. Many retailers now operate formats such as wholesale clubs and hypermarkets that perform many wholesale functions. In return, many large wholesalers are setting up their own retailing operations.

Chapter Outline

1. **Introduction**
   a. Few retailers can compete directly with Wal-Mart. Yet, little Whole Foods is thriving in the shadow of the giant.
   b. Whole Foods succeeds through careful positioning----specifically, by positioning itself away from Wal-Mart. It targets customers that Wal-Mart can't serve, offering them value that Wal-Mart can't deliver.
   c. Whole Foods' value proposition is summed up in its motto: " Whole Foods. Whole People. Whole Planet." In keeping with the company's positioning, most of the store's goods carry labels proclaiming "organic," "100% natural," and "contains no additives."
   d. Whole Foods has found its own very profitable place in the world.

2. **Retailing**
   a. Retailing includes all the activities involved in selling products or services directly to final consumers for their personal, nonbusiness use.
   b. Retailers are businesses whose sales come primarily from retailing.
   c. Nonstore retailing has been growing much faster than has store retailing. Nonstore retailing includes selling to final consumers through direct mail, catalogs, telephone, the Internet, TV home shopping shows, home and office parties, door-to-door contact, vending machines, and other direct selling approaches.

   Types of Retailers
   d. Table 11-1 shows the most important types of retail stores.

1. Retailers can be differentiated on amount of service.
    i.   Self-service retailers serve customers who are willing to perform their own "locate-compare-select" process to save money. Self-service is the basis of all discount operations and is typically used by sellers of convenience goods and nationally branded, fast-moving shopping goods.
    ii.   Limited-service retailers provide more sales assistance because they carry more shopping goods about which customers need information.
    iii. Full-service retailers, such as specialty stores and first-class department stores, offer salespeople who assist customers in every phase of the shopping process.
2. Retailers can also be classified according to the length and breadth of their product assortments.
    i.   Specialty stores carry narrow product lines with deep assortments within those lines.
    ii.   Department stores carry a wide variety of product lines. Service remains the key differentiating factor.
    iii. Supermarkets are the most frequently shopped type of retail store.
    iv.   Convenience stores are small stores that carry a limited line of high-turnover convenience goods.
    v.   Superstores are much larger than regular supermarkets and offer a large assortment of routinely purchased food products, nonfood items, and services.
        a.   Wal-Mart, Kmart, Target, and others offer supercenters, combination food and discount stores that emphasize cross-merchandising.
    vi.   Category killers feature stores the size of airplane hangars that carry a very deep assortment of a particular line with a knowledgeable staff.
    vii. Hypermarkets are huge superstores. Hypermarkets have been very successful in Europe and other world markets, but they have met with little success in the United States.
3. Retailers can be classified according to the prices they charge.
    i.   A discount store sells standard merchandise at lower prices by accepting lower margins and selling higher volume.
    ii.   An off-price retailer buys at less-than-regular wholesale prices and charges consumers less than retail.
        a.   Independent off-price retailers either are owned and run by entrepreneurs or are divisions of larger retail operations.
        b.   Most large off-price retailer operations are owned by bigger retail chains.
    iii. Factory outlets sometimes group together in factory outlet malls and value-retail centers, where dozens of outlet stores

offer prices as low as 50% below retail on a wide range of items.

      iv.   Warehouse clubs (or wholesale clubs or membership warehouses) operate in huge, drafty, warehouselike facilities and offer few frills.

4.   The major types of retail organizations are described in Table 11-2.

      i.   Chain stores are two or more outlets that are commonly owned and controlled.

          a.   A voluntary chain is a wholesaler-sponsored group of independent retailers that engages in group buying and common merchandising.

          b.   A retailer cooperative is a group of independent retailers that bands together to set up a jointly owned, central wholesale operation and conducts joint merchandising and promotion efforts.

      ii.   Franchise systems are normally based on some unique product or service; on a method of doing business; or on the trade name, goodwill, or patent that the franchiser has developed.

      iii.   Merchandising conglomerates are corporations that combine several different retailing forms under central ownership.

Retailer Marketing Decisions

e.   Retailers are always searching for new marketing strategies to attract and hold customers.

f.   Figure 11-1 shows the major marketing decisions retailers face.

1.   Retailers must first define their target markets and then decide how they will position themselves in these markets.

      i.   Too many retailers fail to define their target markets and positions clearly. They try to have "something for everyone" and end up satisfying no market well.

2.   Retailers must decide on three major product variables.

      i.   The retailer's product assortment should differentiate the retailer while matching target shoppers' expectations.

      ii.   The services mix can also help set one retailer apart from another.

      iii.   The store's atmosphere is another element in the reseller's product arsenal.

3.   A retailer's price policy must fit its target market and positioning, product and service assortment, and competition.
   i.   Most retailers seek either high markups on lower volume or low markups on higher volume.
4.   Retailers use any or all of the promotion tools—advertising, personal selling, sales promotion, public relations, and direct marketing—to reach consumers.
5.   Retailers often point to three critical factors in retailing success—location, location, and location.
   i.   It is very important that retailers select locations that are accessible to the target market in areas that are consistent with the retailer's positioning.
   ii.   Central business districts were the main form of retail cluster until the 1950s.
   iii.   A shopping center is a group of retail businesses planned, developed, owned, and managed as a unit.
      a.   A regional shopping center, or regional shopping mall, the largest and most dramatic shopping center, contains from 40 to more than 200 stores.
      b.   A community shopping center contains between 15 and 40 retail stores.
      c.   Most shopping centers are neighborhood shopping centers or strip malls that generally contain between 5 and 15 stores.
      d.   A recent addition to the shopping center scene is the so-called power center. These huge unenclosed shopping centers consist of a long strip of retail stores, each with its own entrance with parking directly in front.

The Future of Retailing

g.   Retailers operate in a harsh and fast-changing environment, which offers threats as well as opportunities.
1.   New retail forms continue to emerge to meet new situations and consumer needs, but the life cycle of new retail forms is getting shorter.
2.   The wheel of retailing concept says that many new types of retailing forms begin as low-margin, low-price, low-status operations. They challenge established retailers that have become "fat" by letting their costs and margins increase. The new retailers' success leads them to upgrade their facilities and offer more services. In turn, their costs increase, forcing them to increase their prices. Eventually, the new retailers become like the conventional retailers they replaced. The cycle begins again.

3. Americans are increasingly avoiding the hassles and crowds at malls by doing more of their shopping by phone or online.
4. Today's retailers are increasingly selling the same products at the same price to the same consumers in competition with a wider variety of other retailers.
    i. This merging of consumers, products, prices, and retailers is called retail convergence.
    ii. This convergence means greater competition for retailers and greater difficulty in differentiating offerings.
5. The rise of huge mass merchandisers and specialty superstores, the formation of vertical marketing systems and buying alliances, and a rash of retail mergers and acquisitions have created a core of superpower megaretailers. They are shifting the balance of power between retailers and producers.
6. Retail technologies are becoming critically important as competitive tools.
    i. Progressive retailers are using advanced information technology and software systems to produce better forecasts, control inventory costs, order electronically from suppliers, send email between stores, and even sell to customers within stores.
7. Retailers with unique formats and strong brand positioning are increasingly moving into other countries.
    i. U.S. retailers are still significantly behind Europe and Asia when it comes to global expansion.
8. There has been a resurgence of establishments that, regardless of the product or service they offer, also provide a place for people to get together.

3. **Wholesaling**
   a. Wholesaling includes all activities involved in selling goods and services to those buying for resale or business use.
   b. Wholesalers are firms engaged primarily in wholesaling activity.
   c. Wholesalers buy mostly from producers and sell mostly to retailers, industrial consumers, and other wholesalers.
   d. Wholesalers add value by performing one or more of the following channel functions:
      1. Selling and promoting
      2. Buying and assortment building
      3. Bulk breaking
      4. Warehousing
      5. Transportation
      6. Financing
      7. Risk bearing
      8. Market information
      9. Management services and advice

<u>Types of Wholesalers</u>

e.   Wholesalers fall into three major groups as shown in Table 11-3.

    1.   Merchant wholesalers are the largest single group of wholesalers, accounting for roughly 50% of all wholesaling.

        i.   Full-service wholesalers provide a full set of services.

        ii.   Limited-service wholesalers offer fewer services to their suppliers and customers.

    2.   Brokers and agents do not take title to goods, and they perform only a few functions. They generally specialize by product line or customer type.

        i.   A broker brings buyers and sellers together and assists in negotiations.

        ii.   Agents represent buyers or sellers on a more permanent basis.

            a.   Manufacturers' agents are the most common type of agent wholesaler.

        iii.   Manufacturers' sales branches and offices are the third major type of wholesaler.

<u>Wholesaler Marketing Decisions</u>

f.   As with retailers, wholesaler marketing decisions include choices of target markets, positioning, and the marketing mix. See Figure 11-2.

    1.   Wholesalers must define their target markets and position themselves effectively. They can choose a target group by size of customer, type of customer, need for service, or other factors.

    2.   Wholesalers must decide on product assortment and services, prices, promotion, and place.

        i.   The wholesaler's "product" is the assortment of products and services that it offers.

        ii.   Price is an important decision. Wholesalers usually mark up the cost of goods by a standard percentage, say 20%.

        iii.   Most wholesalers are not promotion minded.

        iv.   Place is important—wholesalers must choose their locations, facilities, and Web locations carefully.

<u>Trends in Wholesaling</u>

g.   As the wholesaling industry moves into the 21st century, it faces considerable challenges.

h.   The industry remains vulnerable to one of the most enduring trends of the last decade—fierce resistance to price increases and the winnowing out of suppliers who are not adding value based on cost and quality.

i.   The distinction between large retailers and large wholesalers continues to blur.

    1.   Many retailers now operate formats such as wholesale clubs and hypermarkets that perform many wholesale functions.

2. In return, many large wholesalers are setting up their own retailing operations.
j. Wholesalers will continue to increase the services they provide to retailers.
k. Many large wholesalers are now going global.

Creative Marketing Exercises Designed to Reinforce the Concepts!!! (Suggested answers to these exercises can be found at the end of the Study Guide.)

1. Do you prefer to shop in a retail store or in a non-retail store? Justify your answer.

2. Using the three levels of service (self-service, limited-service, and full-service), find 3 examples of each.

3. Visit a local store that has recently renovated its site. Will the changes have a positive effect on the store? Why or why not?

4. Visit www.dollargeneral.com and give several examples of this store as a discount store.

5. Research Target and determine why the company decided to "super size" its stores.

6. Using your favorite search engine, find out all you can about Stanley Tanger. What was his vision and has it come to fruition?

7. Novelty items such as sports fan paraphernalia are a common draw for a small target market. List the pros and cons of adding these items to your inventory.

8. Visit www.ToysRUS.com and describe the atmosphere the website tries to create. Can atmosphere be created successfully?

9. Using your favorite search engine, locate a large mall in your area. How many stores are included in this mall? How many different activities besides shopping are featured?

10. Go to www.morganstanley.com and explain how this company could be considered a wholesaler. What is the firm selling and how?

"Linking the Concepts" (#1) -- Suggestions/Hints

1. Take, for example, a sports fan of the Carolina Panthers professional football team. This person wants to purchase a new sweatshirt with the Carolina Panthers logo on it. He/she could get one of these sweatshirts at the local Wal-Mart or he/she could purchase one at the stadium in one of the concourse (specialty) shops. The sweatshirt at Wal-Mart is going to be cheaper than at the stadium. However, at Wal-Mart, he/she might have a better variety of different teams to choose from. He/she could pick up other things while shopping at Wal-Mart. He/she might be able to get it washed before the game to wear it, too. The fan will be able to write a check for the sweatshirt at Wal-Mart.

If this same fan waited to get the sweatshirt at the stadium, he/she will get the sense of pride that it was purchased at a game, not "generically" at Wal-Mart. This may make them feel like more of a fan than others. There might be a better variety of Carolina Panthers sweatshirts available at the concourse shops. The bad thing about buying the sweatshirt at the stadium is that you won't be able to wash it before your wear it, if you wear it at the game. Also, most shops at the stadium will want either cash or credit cards. Checks will not be taken. The prices of the sweatshirts will be higher, too.

Some people will choose a sweatshirt from Wal-Mart because it will be cheaper. Others will want to purchase it at the stadium, because it "came from the stadium." It was purchased when the fan was supporting his/her team.

2. These two formats of selling sweatshirts will not change in the future. There is more at stake than just prices. There are the psychological aspects of the purchase that impact more at the stadium concourse shops than at Wal-Mart. There is the pride factor of purchasing the sweatshirt at the game and it might be more of an impulse purchase at the stadium. The sweatshirt purchase might not be as much of an impulse item at Wal-Mart. The future of retailing sweatshirts, both at the game and at Wal-Mart, will not change. Different target markets are shopping for that item at those locations. They have different needs that each site is satisfying through their retail mixes.

"Linking the Concepts" (#2) – Suggestions/Hints

1. Let's compare the Barnes & Noble website to a Barnes & Noble store. The website is great in the fact that a consumer can shop around for any book they are looking for, 24 hours a day. A consumer can purchase the books online or find the nearest store to go shopping for a book or magazine. A consumer can do searches for titles, authors, and subjects as examples of some things the website offers.

This is different than if you go to the store. All of those things can be done at a local Barnes & Noble store. However, the consumer will not get the chance to look over the book and compare it to other books of the same genre to determine which appeals more to them. The consumer will miss out on the whole Barnes & Noble experience, too. This includes the ability to buy a cup of coffee, smell the coffee being brewed, sit in an oversized chair and start reading the book, or even getting involved in discussions with other consumers who have similar interests.

Barnes & Noble markets itself as more than just a bookstore and a lot of that "high touch" feeling is lost by using the website. Most consumers of the retail store prefer the shopping experience there. They like to shop for what they need and also shop around for all of the other items that are being sold and what is on sale. Most consumers who like going to the store enjoy the slowed down, relaxed, "library feel" of the Barnes & Noble experience. However, if a consumer needs to find out if a book is at the store and they do not have the time for that retail experience, the website is the best way to shop. They can get online, do a quick search, and then decide if they want to go to the store or order from the site.

2. The Barnes & Noble store is successful at creating the "community feel" that it offers consumers. They have marketed themselves as a whole, quality bookstore experience, and a place to "hang out." This experience includes the shopping, the bakery, the coffee, and the lounge areas. Consumers of Barnes & Noble like and appreciate this product offering. A lot of that "feel" is lost when using the website. The website loses the personal contact and the face-to-face discussions among consumers. It does offer speed and ease for consumers, so there is a need being met by the retail store with the website. Those consumers that want the Barnes & Noble experience will drive to the store for the experience.

Marketing Adventure Exercises (Suggested answers to these exercises can be found at the end of the Study Guide.)

(Visit www.prenhall.com/adventure for advertisements.)

1.   Food   Snickers

Assume that the candy bar shown in the selected ad was sold in a supermarket, convenience store, and wholesale club. Although the actual product would remain the same, how might other marketing variables differ?

2.   Student choice

Review the major store retailer types (Table 11.1) in your text. List and describe them. Find ads for products that would belong in these various categories.

3.   Apparel   Converse
        Auto   Ford
        Food   Pepsi

Discuss the three different ways to classify retailers with regard to the amount of service they provide. Match the selected ads with the most appropriate type of retailer.

4.   Apparel   Hushpuppies, Umbro

Discuss how specialty stores and department stores differ in terms of product line and what changes they have had. Which of the selected ads seems most appropriate for a department store or a specialty store?

5.   Apparel   Levi's

In which type of retail store are Levi's products typically found? What differences might exist between these different stores?

6.   Apparel   Johnsons
       General   Target, Target2

The text states, "Too many retailers fail to define their target markets and positions clearly. They try to have 'something for everyone' and end up satisfying no market well." Look at the stores shown in the selected ads and determine which is better at defining its target market.

7.   Student choice

Discuss the differences between regional shopping centers, community shopping centers, and neighborhood shopping centers. Peruse all the ads and select products that would be carried in retail stores in a typical neighborhood shopping center.

8.   Electronics   Energizer

Consider the product in the selected ad. Discuss all the retailers that could carry the item.

9.   Student choice

What is retail convergence? Select an ad and discuss that product in terms of this trend.

10.   Cosmetics   Aim
       Household   Paloma

Review the different types of wholesalers. Match the selected products to the appropriate type of wholesaler.

SUM IT UP!!!!!!

Using only this page, sum up all of the concepts and terms discussed in Chapter 11 – "Retailing and Wholesaling". Here is your chance to make sure you know and understand the concepts!!!!

Chapter 12
Communicating Customer Value:
Advertising, Sales Promotion, and Public Relations

Previewing the Concepts: Chapter Objectives

1.   Discuss the process and advantages of integrated marketing communications in communicating customer value.
2.   Define the five promotion tools and discuss the factors that must be considered in shaping the overall promotion mix.
3.   Describe and discuss the major decisions involved in developing an advertising program.
4.   Explain how sales promotion campaigns are developed and implemented.
5.   Explain how companies use public relations to communicate with their publics.

JUST THE BASICS

Chapter Overview

This chapter discusses the importance of coordinating the company's marketing mix components and integrating all the messaging elements into one cohesive unit. A company's marketing communications mix consists of a specific blend of advertising, sales promotion, public relations, personal selling, and direct-marketing tools that the company uses to pursue its advertising and marketing objectives.

Customers don't distinguish between message sources the way marketers do. In the consumer's mind, advertising messages from different media and different promotional approaches all become part of a single message about the company. Conflicting messages from these different sources can result in confused company images and brand positions. The problem is that different communications usually come from different company sources. Under integrated marketing communications, the company carefully integrates and coordinates its many communications channels to deliver a clear, consistent, and compelling message about the organization and its brands.

Integrated marketing communications involve identifying the target audience and shaping a well-coordinated promotional program to elicit the desired audience response. Marketers are moving toward viewing communications as managing the customer relationship over time. Thus, the communications process should start with an audit of all the potential contacts target customers may have with the company and its brands.

The chapter covers each marketing communications and promotion tool in detail, explaining how they are each used, their advantages and disadvantages, and how to best plan each communication, as well as the overall process and communications

The chapter covers each marketing communications and promotion tool in detail, ex-plaining how they are each used, their advantages and disadvantages, and how best to plan each communication as well as the overall process and communications mix. Also described are the factors that influence the marketer's choice of promotion tools.

Chapter Outline

1. **Introduction**
   a. Crispin Porter + Bogusky (CP+B) is right where it's at in today's advertising. Working with modest ad budgets, CP + B has riveted customers' attention with startling guerrilla tactics, unconventional use of media, and holistic marketing strategies that tie everything together.
   b. The term "integrated marketing communication" describes CP + B's approach. Appropriate adjectives include: fresh, radical, street-smart, mischievous, all-over-the-lot, maybe-the-next-best-thing.
   c. During the 1990s, CP+B won local awards, but in 1997, the firm drew national attention through the teen antismoking campaign "Truth." Based on street-level research with local teenagers, CP+B used guerilla-ambush tactics to create an "anti-brand"---"Truth."
   d. Between 1998 and 2002, smoking among middle and high school students in Florida declined an average of 38 percent.
   e. "Truth" begat the celebrated BMW MINI campaign using inexpensive, offbeat communications. This has become one of the most celebrated marketing efforts in recent years, scooping up numerous advertising industry awards.
   f. Building good customer relationships calls for more than just developing a good product, pricing it attractively, and making it available to customers.
      1. Companies must also communicate the value to customers, and what they communicate should not be left to chance.
      2. All of their communications efforts must be blended into a consistent and coordinated communications program.

2. **The Promotion Mix**
   a. A company's total marketing communications mix—also called its promotion mix—consists of the specific blend of advertising, sales promotion, public relations, personal selling, and direct-marketing tools that the company uses to persuasively communicate customer value and build customer relationships. Definitions of the five major promotion tools follow:
      1. Advertising is any paid form of nonpersonal presentation and promotion of ideas, goods, or services by an identified sponsor.

2.   Sales promotions are short-term incentives to encourage the purchase or sale of a product or service.
3.   Public relations involves building good relations with the company's various publics by obtaining favorable publicity, building up a good corporate image, and handling or heading off unfavorable rumors, stories, and events.
4.   Personal selling is personal presentation by the firm's sales force for the purpose of making sales and building customer relation-ships.
5.   Direct marketing establishes direct connections with carefully targeted individual consumers to both obtain an immediate response and cultivate lasting customer relationships.

b.   Communication goes beyond these specific promotion tools.
1.   The product's design, its price, the shape and color of its package, and the stores that sell it—all communicate something to buyers.
2.   Although the promotion mix is the company's primary communication activity, the entire marketing mix—promotion and product, price, and place—must be coordinated for the greatest communication impact.

3.   **Integrated Marketing Communications**
a.   During the past several decades, companies around the world have perfected the art of mass marketing.
b.   However, as we move into the twenty-first century, marketing managers face some new marketing communications realities.

The New Marketing Communications Landscape
c.   Two major factors are changing marketing communications.
1.   marketers are shifting away from mass marketing
2.   improvements in information technology are speeding the movement towards segmented marketing.
d.   Improved information technology has caused striking changes in the ways in which companies and customer communicate with one another.

The Shifting Marketing Communications Model
e.   The shift from mass marketing to segmented marketing has had a dramatic impact on marketing communications. The shift toward targeted marketing and the changing communications environment are giving birth to a new marketing communications model.
f.   Companies are doing less broadcasting and more narrowcasting. Fragmentation will become the opportunity to reach----small clusters of consumers who are consuming exactly what they want.

g. The new communications model will consist of a shifting mix of traditional mass media and a wide array of targeted, personalized media.

The Need for Integrated Marketing Communications

h. Customers don't distinguish between message sources the way marketers do. In the consumer's mind, advertising messages from different media and different promotional approaches all become part of a single message about the company. Conflicting messages from these different sources can result in confused company images, brand positions, and customer relationships.

i. Companies often fail to integrate their various communications channels. Mass-media advertisements say one thing, a price promotion sends a different signal, a product label creates still another message, company sales literature says something altogether different, and the company's website seems out of sync with everything else.

j. The problem is that these communications often come from different company sources.

k. Under the concept of integrated marketing communications, the company carefully integrates and coordinates its many communications channels to deliver a clear, consistent, and compelling message about the organization and its brands. The concept is illustrated in Figure 12-1.

l. IMC calls for recognizing all contact points where the customer may encounter the company, its products, and its brands. Each brand contact will deliver a message—whether good, bad, or indifferent. The company must strive to deliver a consistent and positive message with each contact.

m. IMC ties together all of the company's messages and images.

n. Integrated marketing communications involves identifying the target audience and shaping a well-coordinated promotional campaign to elicit the desired audience response.

o. Marketers are moving toward viewing communications as managing the customer relationship over time. Because customers differ, communications programs need to be developed for specific segments, niches, and even individuals.

p. In the days of new interactive digital communications technologies, companies must ask not only, "How can we reach our customers?" but also, "How can we find ways to let our customers reach us?"

4. **Shaping the Overall Promotion Mix**
a. The concept of integrated marketing communications suggests that the company must blend the promotion tools carefully into a coordinated promotion mix.

161

b. The factors that influence the marketer's choice of promotion tools follow. Each promotion tool has unique characteristics and costs.
  1. Advertising can reach masses of geographically dispersed buyers at a low cost per exposure, and it enables the seller to repeat a message many times.
     i. Large-scale advertising says something positive about the seller's size, popularity, and success.
     ii. Consumers tend to view advertised products as more legitimate.
     iii. Advertising also has shortcomings. Although it reaches people quickly, advertising is impersonal. It can carry on only a one-way communication with the audience, and the audience does not feel that it has to pay attention or respond. It can be very costly.
  2. Personal selling is the most effective tool at certain stages of the buying process, particularly in building up buyers' preferences, convictions, and actions.
     i. It involves personal interaction between two or more people, so each person can observe the other's needs and characteristics and make quick adjustments.
     ii. The effective salesperson keeps the customer's interests at heart in order to build a long-term relationship.
     iii. A sales force requires a longer-term commitment than does advertising. Personal selling is the company's most expensive promotion tool.
  3. Sales promotion includes a wide assortment of tools. These tools attract consumer attention, offer strong incentives to purchase, and can be used to dramatize product offers and to boost sagging sales.
  4. Public relations is very believable.
     i. Public relations can reach many prospects who avoid salespeople and advertisement.
     ii. A well-thought-out public relations campaign used with other promotion mix elements can be very effective and economical.
  5. Direct marketing is nonpublic. The message is normally directed to a specific person.
     i. It is immediate and customized.
     ii. It is interactive.
c. Marketers can choose from two basic promotion mix strategies. See Figure 12-2.
  1. A push strategy involves "pushing" the product through distribution channels to final consumers. The producer directs its marketing activities toward channel members to induce them to carry the product and to promote it to final consumers.

2. In a pull strategy, the producer directs its marketing activities toward final consumers to induce them to buy the product. Consumers will demand the product from channel members, who will in turn demand it from producers.

d. Companies consider many factors when designing their promotion mix strategies, including type of product/market and the product life-cycle stage.

5. Advertising

a. Advertising can be traced back to the very beginnings of recorded history.

b. Although advertising is used mostly by business firms, it is also used by a wide range of not-for-profit organizations, professionals, and social agencies that advertise their causes to various target publics.

c. Advertising is a good way to inform and persuade.

d. Marketing management must make four important decisions when developing advertising campaign. See Figure 12-3.

Setting Advertising Objectives

e. The first step is to set advertising objectives. These objectives should be based on past decisions about the target market, positioning, and marketing mix, which define the job that advertising must do in the total marketing program.

f. An advertising objective is a specific communication task to be accom-plished with a specific target audience during a specific period of time.

1. Advertising objectives can be classified by primary purpose—whether the aim is to inform, persuade, or remind. Table 12-1 lists examples.

   i. Informative advertising is used heavily when introducing a new product category. In this case, the objective is to build primary demand.

   ii. Persuasive advertising becomes more important as competition increases. Here, the company's objective is to build selective demand. Some persuasive advertising becomes comparative advertising, in which a company directly or indirectly compares its brand with one or more other brands.

   iii. Reminder advertising is important for mature products—it keeps consumers thinking about the product.

Setting the Advertising Budget

g. After determining its advertising objectives, the company next sets its advertising budget for each product.

h. There are four common methods used to set the total budget for advertising.
1. In the affordable method, the company sets the promotion budget at the level it thinks it can afford.
   i. Small businesses often use this method.
   ii. This method of setting budgets completely ignores the effects of promotion on sales.
      a. It places advertising last among spending priorities.
      b. It leads to an uncertain annual promotion budget.
2. In the percentage-of-sales method, the company sets the promotion budget at a certain percentage of current or forecasted sales.
   i. There are advantages, because it is simple to use and helps management think about the relationships between promotion spending, selling price, and profit per unit.
   ii. But this method has little to justify it.
      a. It wrongly views sales as the cause of promotion rather than as the result.
      b. It is based on availability of funds rather than opportunities.
      c. It may prevent increased spending that is sometimes needed to turn around falling sales.
      d. Because the budget varies with year-to-year sales, long-range planning is difficult.
      e. The method does not provide any basis for choosing a specific percentage, except what has been done in the past or what competitors are doing.
3. In the competitive-parity method, companies set their promotion budget to match competitors' outlays.
   i. Competitors' budgets represent the collective wisdom of the industry. Spending what competitors spend helps prevent promotion wars.
   ii. But there are no grounds for believing that the competition has a better idea of what a company should be spending on promotion. And there is no evidence that budgets based on competitive parity really do prevent promotion wars.
4. The most logical budget-setting method is the objective-and-task method. Here, the company sets its promotion budget based on what it wants to accomplish with promotion.
   i. This method entails defining specific promotion objectives, determining the tasks needed to achieve these objectives, and estimating the costs of performing these tasks. The sum of these costs is the proposed promotion budget.
   ii. This method forces management to spell out its assumptions about the relationship between dollars spent and pro-motion results.

        iii. It is also the most difficult method to use.

Developing Advertising Strategy
i.   Advertising strategy consists of two major elements: creating advertising messages and selecting advertising media.
j.   In the past, companies often viewed media planning secondary to the message-creation process. But media fragmentation, soaring media costs, and more-focused target marketing strategies have promoted the importance of the media-planning function.
k.   Good advertising messages are especially important in today's costly and cluttered advertising environment.
l.   With the growth in cable and satellite TV, VCRs, and remote-control units, today's viewers have many more options. They can avoid ads by watching commercial-free cable channels. They can "zap" commercials by pushing the fast-forward button during taped programs.
      1.   Many advertisers now see themselves as creating "advertainment"—ads that are both persuasive and entertaining.
m.   The first step in creating effective advertising messages is to plan a message strategy—to decide what general message will be communicated to consumers.
      1.   Developing an effective message strategy begins with identifying customer benefits that can be used as advertising appeals. Ideally, advertising message strategy will follow directly from the company's broader positioning strategy.
      2.   The next step is to develop a compelling creative concept or "big idea" that will bring the message strategy to life in a distinctive and memorable way.
      3.   The creative concept will guide the choice of specific appeals to be used. Advertising appeals should have three characteristics: They should be meaningful, pointing out benefits that make the product more desirable or interesting to consumers; they should be believable; and they should be distinctive.
      4.   The advertiser now has to turn the big idea into an actual ad execution that will capture the target market's attention and interest. Any message can be presented in different execution styles
            i.   Slice of life: shows one or more "typical" people using the product in a normal setting.
            ii.   Lifestyle: shows how a product fits in with a particular lifestyle.
           iii.   Fantasy: creates a fantasy around the product or its use.
           iv.   Mood or image: builds a mood or image around the product, such as beauty, love, or serenity.
            v.   Musical: shows one or more people or cartoon characters singing about the product.

      vi.   Personality symbol: creates a character that represents the product.

     vii.   Technical expertise: shows the company's expertise in making the product.

   viii.   Scientific evidence: presents survey or scientific evidence that the brand is better or better liked than one or more other brands.

     ix.   Testimonial evidence or endorsement: features a highly believable or likable source endorsing the product.

5. The advertiser must also choose a tone for the ad.
6. The advertiser must use memorable and attention-getting words in the ad.
7. Format elements make a difference in an ad's impact as well as its cost.
8. The illustration is the first thing the reader notices.
9. The headline must effectively entice the right people to read the copy.

      i.   The copy—the main block of text in the ad—must be simple but strong and convincing.

     ii.   These three elements must work together effectively.

n. The major steps in media selection are deciding on reach, frequency, and impact; choosing among major media types; selecting specific media vehicles; and deciding on media timing.

1. Reach is a measure of the percentage of people in the target market who are exposed to the ad campaign during a given period of time. Frequency is a measure of how many times the average person in the target market is exposed to the message. Media impact is the qualitative value of a message exposure through a given medium.

2. The media planner has to know the reach, frequency, and impact of each of the major media types. Table 12-2 summarizes the media types.

      i.   Media planners consider many factors when making their media choices.

         a.   The target consumers.

         b.   The medium's impact.

         c.   The message effectiveness.

         d.   Cost is another major factor. Media impact and cost must be reexamined regularly.

3. The media planner must now choose the best media vehicles—specific media within each general media type.

      i.   Media planners must compute the cost per thousand persons reach by a vehicle.

     ii.   The media planner must also consider the costs of producing ads for the different media.

    iii.   In selecting media vehicles, the media planner must balance media cost measures against several media effectiveness factors: audience quality, audience attention, and editorial quality.

4. The advertiser must also decide how to schedule the advertising over the course of a year. The advertiser also has to choose the pattern of the ads.

    i. Continuity means scheduling ads evenly within a given period.

    ii. Pulsing means scheduling ads unevenly over a given time period.

Evaluating Advertising

o. The advertising program should regularly evaluate both the communication effects and the sales effects of advertising.

1. Measuring the communication effects of an ad or ad campaign tells whether the ad is communicating well. Copy testing can be done before or after an ad is printed or broadcast.

2. The sales and profits effects of advertising are often harder to measure than the communication effects. Sales are affected by many factors besides advertising, such as product features, price, and availability.

    i. One way to measure sales and profits effects of advertising is to compare past sales and profits with past advertising expenditures.

    ii. Another way is through experiments such as varying the spending in different markets.

Other Advertising Considerations

p. The company must address two additional questions. First, how will the company organize its advertising function? Second, how will the company adapt its advertising strategies and programs to the complexities of inter-national markets?

1. Different companies organize in different ways to handle advertising.

    i. In small companies, advertising might be handled by someone in the sales department.

    ii. Large companies set up advertising departments whose job it is to set the advertising budget, work with the ad agency, and handle advertising not done by the agency.

    iii. Advertising agencies employ specialists who can often perform advertising tasks better than can the company's own staff.

        a. Most large advertising agencies have the staff and resources to handle all phases of an advertising campaign for its clients, from creating a marketing plan to developing ad campaigns and preparing, placing, and evaluating ads.

2. International advertisers face many complexities not encountered by domestic advertisers.
   i. The most basic issue concerns the degree to which global advertising should be adapted to the unique characteristics of markets in various countries.
      a. Standardization produces many benefits—lower advertising costs, greater global advertising coordination, and a more consistent worldwide image.
      b. There are also drawbacks. It ignores the facts that country markets differ greatly in their cultures, demographics, and economic conditions.
   ii. Global advertisers face several special problems.
      a. Advertising media costs and availability differ vastly from country to country.
      b. Countries also differ in the extent to which they regulate advertising practices.
   iii. Although advertisers may develop global strategies to guide their overall advertising efforts, specific advertising programs must usually be adapted to meet local cultures and customers, media characteristics, and advertising regulations.

6. Sales Promotion

a. Sales promotion consists of short-term incentives to encourage the purchase or sales of a product or service.
b. Whereas advertising and personal selling offer reasons to buy a product or service, sales promotion offers reasons to buy now.

Rapid Growth of Sales Promotion

c. Sales promotion tools are used by most organizations, including manufacturers, distributors, retailers, trade associations, and not-for-profit institutions.
d. They are targeted toward final buyers, retailers, and wholesalers; business customers; and members of the sales force.
e. Several factors have contributed to the rapid growth of sales promotions.
   1. Inside the company, product managers face greater pressures to increase their current sales, and promotion is viewed as an effective short-run sales tool.
   2. Externally, the company faces more competition and competing brands are less differentiated.
   3. Advertising efficiency has declined because of rising costs, media clutter, and legal restraints.
   4. Consumers have become more deal-oriented, and ever-larger retailers are demanding more deals from manufacturers.
f. The growing use of sales promotion has resulted in promotion clutter, similar to advertising clutter.

<u>Sales Promotion Objectives</u>

g. Sales promotion objectives vary widely.
   1. Sellers may use consumer promotions to increase short-term sales or to help build long-term market share.
   2. Objectives for trade promotions include getting retailers to carry new items and more inventory, getting them to advertise the product and give it more shelf space, and getting them to buy ahead.
   3. For the sales force, objectives include getting more sales force support for current or new products or getting salespeople to sign up new accounts.
   4. Sales promotion should reinforce the product's position and build long-term customer relationships.

<u>Major Sales Promotion Tools</u>

h. Many tools can be used to accomplish sales promotion objectives.
   1. The main consumer promotion tools include those following:
      i. Samples are offers of a trial amount of a product. Sampling is the most effective, but most expensive, way to introduce a new product.
      ii. Coupons are certificates that give buyers savings when they purchase specified products.
      iii. Cash refund offers (or rebates) are like coupons except that the price reduction occurs after the purchase rather than at the retail outlet. The consumer sends a "proof of purchase" to the manufacturer, who then refunds part of the purchase price by mail.
      iv. Price packs (also called cents-off deals) offer consumers savings off the regular price of a product. The reduced prices are marked by the producer directly on the label or package.
      v. Premiums are goods offered either free or at low cost as an incentive to buy a product. A premium may come inside the package (in-pack), outside the package (on-pack), or through the mail.
      vi. Advertising specialties, also called promotional products, are useful articles imprinted with an advertiser's name that are given as gifts to consumers.
      vii. Patronage rewards are cash or other awards offered for the regular use of a certain company's products or services.
      viii. Point-of-purchase (POP) promotions include displays and demonstrations that take place at the point of purchase or sale.
      ix. Contests, sweepstakes, and games give consumers the chance to win something.

       a.   A contest calls for consumers to submit an entry to be judged by a panel that will select the best entries.

       b.   A game sweepstakes calls for consumers to submit their names for a drawing.

       c.   A game presents consumers with something every time they buy, which may or may not help them win a prize.

2.   Manufacturers direct more sales promotion dollars toward retailers and wholesalers (78%) than to consumers (22%).

    i.   Trade promotion can persuade resellers to carry a brand, give it shelf space, promote it in advertising, and push it to consumers.

    ii.   Manufacturers have several trade promotion tools.

       a.   Many of the tools used for consumer promotions, such as contests, premiums, and displays, can also be used as trade promotions.

       b.   A discount is off the list price on each case purchased during a stated period of time (also called a price-off, off-invoice, or off-list).

       c.   An allowance can be offered in return for the retailer's agreement to feature the manufacturer's products in some way.

          i.   An advertising allowance compensates retailers for advertising the product.

          ii.   A display allowance compensates retailers for using special displays.

       d.   Manufacturers may offer free goods, which are extra cases of merchandise, to resellers who buy a certain quantity or feature a certain flavor or size.

       e.   Manufacturers may offer push money—cash or gifts to dealers or their sales forces—to "push" the manufacturer's goods.

       f.   Manufacturers may give retailers free specialty advertising items that carry the company's name.

3.   Business promotion tools are used to generate business leads, stimulate purchases, reward customers, and motivate salespeople.

    i.   Business promotion tools include many of the same tools used for consumer or trade promotion.

    ii.   Many companies and trade associations organize conventions and trade shows to promote their products. Firms selling to the industry show their products at the trade show.

    iii.   A sales contest is a contest for salespeople or dealers to motivate them to increase their sales performance over a given period.

Developing the Sales Promotion Program
i.   The marketer must make several other decisions in order to define the full sales promotion program.
   1.   The marketer must decide on the size of the incentive.
   2.   The marketer must set conditions for participation.
   3.   The marketer must decide how to promote and distribute the promotion program itself.
   4.   The marketer must consider the length of the promotion, which is also important.
   5.   The marketer must also evaluate the program.
j.   The most common evaluation method is to compare sales before, during, and after a promotion.
   1.   Marketers should ask if the promotion attracted new customers or more purchasing from current customers.
   2.   It is important to ask if the company hold on to these new customers and purchases?
   3.   Marketers should try to determine if the long-run customer relationship and sales gains from the promotion justify its costs?

7.   Public Relations
   a.   Public relations is building good relations with the company's various publics by obtaining favorable publicity, building up a good corporate image, and handling or heading off unfavorable rumors, stories, and events.
   b.   Public relations departments may perform any or all of the following functions:
      1.   Press relations or press agency.
      2.   Product publicity.
      3.   Public affairs.
      4.   Lobbying.
      5.   Investor relations.
      6.   Development.
   c.   Public relations is used to promote products, people, places, ideas, activities, organizations, and even nations.

<u>The Role an Impact of Public Relations</u>
   d.   Public relations can have a strong impact on public awareness at a much lower cost than advertising can.
   e.   The company does not pay for the space or time in the media. Rather, it pays for a staff to develop and circulate information and to manage events.
   f.   Public relations is often described as a marketing stepchild because of its limited and scattered use. However PR is playing an increasingly important brand-building role.

<u>Major Public Relations Tools</u>

g. Public relations professionals use several tools.

    1.   One of the major tools is news. PR professionals find or create favorable news about the company and its products or people.

    2.   Speeches can also create product and company publicity.

    3.   Special events range from news conferences, press tours, grand openings, and fun events.

    4.   Written materials including annual reports, brochures, articles, and company newsletters and magazines are often produced.

    5.   Audiovisual materials, such as films, are being used increasingly as communication tools.

    6.   Corporate identify materials, such as logos, stationery, brochures, signs, business forms, business cards, buildings, uniforms, and company cars and trucks, all become marketing materials.

    7.   Public service activities can improve public goodwill.

    8.   Buzz marketing campaigns generate publicity by getting consumers themselves to spread information about a product or service to others in their communities.

    9.   Another development is Mobile marketing---traveling promotional tours that bring the brand to consumers.

    10.   A company's website can be a good public relations vehicle. Websites are also ideal for handling crisis situations by disseminating information through e-mail marketing, online chat, and blogs. Public relations is a valuable part of doing business in a digital world.

h. As with other promotion tools, management should set PR objectives, choose the PR messages and vehicles, implement the PR plan, and evaluate the results.

    i.   The company's public relations should be blended smoothly with other promotion activities within the company's overall integrated marketing communications efforts.

<u>Creative Marketing Exercises Designed to Reinforce the Concepts!!! (Suggested answers to these exercises can be found at the end of the Study Guide.)</u>

    1.   Find 10 examples of websites that use color and animation to attract children to their products.

    2.   Visit www.bandaid.com and show how Johnson & Johnson sells Band Aids to a large variety of customers.

    3.   Outline the 9 elements of the communication process in 5 30-second commercials. How do the different companies relay their messages differently?

    4.   Give 5 examples of products that are more effectively marketed in a print ad than on the radio.

5.	Pretend you are going to buy a new flat screen plasma television set. Outline your internal dialogue through each of the 6 stages of buyer readiness.

6.	Locate 5 websites that use the "slice of life" style in their advertising. Justify your selections.

7.	Why is the Super Bowl such an attractive time slot to premier new commercials? What is the average cost of a 30-second time slot during halftime and is this cost justified?

8.	Identify a company that uses a celebrity to sell its product. Do a brief biographical sketch of this person.

9.	Visit your local cosmetics counter and inquire about their "gift with purchase" sales incentives. Does the salesperson see a marked jump in sales during that time?

10.	Recall a human-interest story that prompted you to buy something or make a contribution. Why was this story different from the countless others we are exposed to?

"Linking the Concepts" (#1) – Suggestions/Hints

1.	Companies use integrated marketing communications (IMC) to communicate customer value. Modern marketing calls for more than just creating customer value by developing a good product, pricing it attractively, and making it available to target customers. Companies also must clearly and persuasive communicate that value to current and prospective customers. Guided by an overall IMC strategy, the company works out the roles that the various promotional tools will play and the extent to which each will be used. It carefully coordinates the promotional activities and the timing of when major campaigns take place.

2.	Recent shifts toward targeted or one-to-one marketing, coupled with advances in information and communication technology, have had a dramatic impact on marketing communications. The digital age has spawned a host of new information and communication tools—from the Internet and cell phone to satellite and cable television systems and digital video recorders (DVRs). The new technologies give companies new digital tools for interacting with targeted customers. They also give consumers more control over the nature and timing of messages they choose to receive.

	Consumers that are in the market for a new car might hear about various available models on television, radio, or on the Internet. Consumers have a world of information available to them via the Internet. For example, www.edmunds.com provides consumers with just about any and all information (safety ratings, pricing, etc.) they need in order to make a well-informed purchase decision.

"Linking the Concepts" (#2) – Suggestions/Hints

1. A good example of an interesting advertisement would be a recently run ad for a local specialty store called Field's of Dreams located in Spartanburg, South Carolina. It is a store that specializes in very fancy gifts. They had a print ad that was trying to generate the interest of busy business people to swing by the store and pick up a quality gift for someone special for Christmas. It was laid out in a "menu" appearance format. It was very effective because it highlighted quality items like Vera Bradley handbags, and yet it was done in a professional manner that respected the time constraints of the business consumer. It was done in a classy format, and included a small price list that highlighted a few items that could be purchased. The color scheme was done very tastefully. It was effective because it maintained a very elegant and professional image in a very generic format (print advertising).

   An example of commercials that people may like, but are not effective, are some political ads. In some local areas, some political ads are very entertaining. The people of the community may talk about the ads. The ad may generate a lot of interest and news; however, there is still a decline in voting across the country. People see entertaining ads, but they are not moved enough to vote for the candidate.

2. The objectives of the Field's of Dreams print advertisements were to make busy, time constrained business people aware of their store and what they had to offer as Christmas was nearing. The "menu" format of the ad was supposed to help the consumer save time by having this quick look at items and prices. The budget was small because the owner has only been in business for a short period of time. He worked on the wording and layout himself, to keep the costs down. It was very effective because it did generate sales and the advertisement maintained the elegant image the store is trying to project. The strategy of presenting the store in an elegant and professional image was important to the owner.

Marketing Adventure Exercises (Suggested answers to these exercises can be found at the end of the Study Guide.)

(Visit www.prenhall.com/adventure for advertisements.)

1. Auto   Chevy, Chevy2

Define advertising. Discuss the advantages and disadvantages of advertising. Look at the two selected ads and decide which builds up a long-term image for a product and which is designed to encourage quick sales of the specific product.

2.   Electronics   Student choice

With integrated marketing communications, the organization coordinates its communications to deliver a clear and consistent message about the organization and its brands. Select an ad in the electronics category and visit that company's website. Discuss whether you believe the message in the ad and that from the website are clear and consistent.

3.   Food   Perdue3

Discuss the two basic promotion mix strategies. Which strategy is used for the selected ad?

4.   Food   Ecusa, Gatorade, Pepsi2, San Luis, San Vicente, Tabasco

Advertising can inform, persuade, or remind. Which ads would be typical of each of these objectives? Match the selected ads in the food category with each of these goals and explain your reasoning.

5.   Financial   Student choice

Advertising can inform the public about a company's products, or the organization itself. Select an ad in the financial category that promotes an organization, not a specific product.

6.   Student choice

Your text describes various types of message execution. List and describe these execution styles. Wherever possible, match ads you select to these styles.

7.   Nonprofit   Eye donor3, Greenpeace

Deciding on the tone for an ad is important. Discuss the "tone" of the selected ads.

8.   Electronics   Sony, Sony2

What are the components to consider when deciding on the format of an advertisement? Discuss the format of the selected ads.

9.   Apparel   Johnsons

Different types of messages may require different media. Which media type is most appropriate for the message in the selected ad?

10.   Student choice

List and describe the main consumer promotion tools.   Find an example of consumer sales promotion incorporated within an advertisement.

11.   Student choice

Discuss the elements included in sales promotion, distinguishing between those aimed at the consumer and non-consumer markets.   Select one ad and convert the message to make it include an element of sales promotion.

12.   Student choice

Define public relations and discuss the various functions of a public relations department. Which category has the most examples of public relations?

Using only this page, sum up all of the concepts and terms discussed in Chapter 12 – "Communicating Customer Value: Advertising, Sales, Promotion, and Public Relations". Here is your chance to make sure you know and understand the concepts!!!!

Chapter 13
Communicating Customer Value:
Personal Selling and Direct Marketing

Previewing the Concepts: Chapter Objectives

1. Discuss the role of a company's salespeople in creating value for customers and building customer relationships.
2. Identify and explain the six major sales force management steps.
3. Discuss the personal selling process, distinguishing between transaction-oriented marketing and relationship marketing.
4. Define direct marketing and discuss its benefits to customers and companies.
5. Identify and discuss the major forms of direct marketing.

JUST THE BASICS

Chapter Overview

This chapter continues the discussion of communication methods begun in Chapter 12. It focuses on personal selling and direct marketing. Personal selling is the interpersonal arm of marketing communications in which the sales force interacts with customers and prospects to make sales and build relationships. Direct marketing consists of direct connections with carefully targeted consumers to both obtain an immediate response and cultivate lasting customer relationships.

Selling is one of the oldest professions in the world. Today, most salespeople are well-educated, well-trained professionals who work to build and maintain longer-term customer relationships. They listen to their customers, assess customer needs, and organize the company's efforts to solve customer problems. The sales force serves as a critical link between a company and its customers.

A sales force can be organized such that it has a territorial structure. In this structure, each salesperson is assigned to an exclusive geographic area and sells the company's full line of products or services to all customers in that territory. A product sales force structure is one in which the sales force sells along product lines. In a customer sales force structure, the sales force is organized along customer or industry lines. Many companies, particularly those that sell a wide variety of products to many types of customers over a broad geographic area, use a complex sales force structure that combines several types.

Personal selling consists of a seven-step process. The first is prospecting and qualifying, followed by the preapproach, approach, presentation, handling objections, closing, and follow-up. All of this should lead to long-term customer relationships.

Direct marketing consists of direct connections with carefully targeted individual consumers to both obtain an immediate response and cultivate lasting customer relationships. Most companies still use direct marketing as a supplementary channel or medium for marketing their goods. However, for many companies today, direct marketing is more than that—it constitutes a new and complete model for doing business.

Effective direct marketing begins with a good customer database. This database is an organized collection of comprehensive data about individual customers or prospects, including geographic, demographic, psychographic, and behavioral data. The database can be used to locate good potential customers, tailor products and services to the special needs of targeted consumers, and maintain long-term customer relationships.

There are several forms of direct marketing, including telephone marketing, direct mail marketing, catalog marketing, direct response television marketing, and kiosk marketing. A very powerful approach for many companies is integrated direct marketing, which involves using carefully coordinated multiple-media, multiple-stage campaigns.

Direct marketers and their customers usually enjoy mutually rewarding relationships. Sometimes, however, a darker side emerges. The aggressive and sometimes shady tactics of a few direct marketers can bother or harm consumers, giving the entire industry a black eye. Direct marketers know that, left untended, such problems will lead to increasingly negative consumer attitudes, lower response rates, and calls for more restrictive state and federal legislation.

Chapter Outline

1. **Introduction**
    a. Perhaps no industry felt the recent economic slowdown more than the technology sector—total information technology spending has been flat for several years. But despite the slump, CDW Corporation, the nation's largest reseller of technology products and services, is thriving.
    b. The company owes its success to its highly effective "clicks and people" direct marketing strategy. CDW's direct model combines good old-fashioned high-touch personal selling with a modern high-tech Web presence to build lasting one-to-one customer relationships.
    c. Whereas many of CDW's competitors chase after a relative handful of very large customers, CDW has traditionally targeted small and midsize businesses (SMBs). These smaller customers often need lots of advice and support.
    d. The major responsibility for building and managing customer relationships falls to CDW's sales force of nearly 1,880 account managers. Each customer is assigned an account manager, who helps the customer select the right products and technologies and keep them running smoothly.

e. Account managers do more than just sell technology products and services. They work closely with customers to find solutions to their technology problems.

f. Customers who want to access CDW's products and expertise without going through their account manager can do so easily at any of several CDW Web sites.

g. When someone says "salesperson" you may still think of the stereotypical "traveling salesman"—the fast-talking, ever-smiling peddler who travels his territory foisting his wares on reluctant customers. Such stereotypes, however, are sadly out of date. Today, like CDW's account managers, most professional salespeople are well-educated, well-trained men and women who work to build valued customer relationships.

2. Personal Selling

a. Sales forces are found not only in business organizations that sell products and services, but also in many other kinds of organizations.

1. Colleges use recruiters to attract new students.
2. Churches use membership committees to attract new members.
3. Hospitals and museums use fund-raisers to contact donors and raise money.
4. The U.S. Postal Service uses a sales force to sell Express Mail and other services to corporate customers.

<u>The Nature of Personal Selling</u>

b. The people who do the selling go by many names: salespeople, sales representative, account executive, sales consultants, sales engineers, agents, district managers, marketing representatives, and account development reps are a few of the names.

c. The term *salesperson* covers a wide range of positions.

1. At one extreme, a salesperson might be largely an order taker, such as the department store salesperson standing behind the counter.
2. At the other extreme are order getters, whose positions demand the creative selling of products and services.

<u>The Role of the Sales Force</u>

d. Personal selling is the interpersonal arm of the promotion mix. It involves two-way, personal communication between salespeople and individual customers.

1. It can be more effective than advertising in more complex selling situations.
2. The sales force serves as a critical link between a company and its customers.
   i. They represent the company to customers.
   ii. The salespeople also represent customers to the company, acting inside the firm as champions of customers' interests and managing the buyer-seller relationship.

3. Salespeople need to be concerned with more than just producing sales—they should work with others in the company to produce customer value and company profit.

3. Managing the Sales Force
a. Sales force management is the analysis, planning, implementation, and control of sales force activities. The major sales force management decisions are shown in Figure 13-1.

<u>Designing Sales Force Strategy and Structure</u>
b. A company can divide sales responsibilities along any of several lines.
    1. In the territorial sales force structure, each salesperson is assigned to an exclusive geographic area and sells the company's full line of products or services to all customers in that territory.
        i. This organization clearly defines each salesperson's job and fixes accountability.
        ii. This sales method increases the salesperson's desire to build local business relationships that improve selling effectiveness.
        iii. This type of organization is often supported by many levels of sales management positions.
    2. In the product sales force structure, the sales force sells along product lines.
        i. The product structure can lead to problems if a single large customer buys many different company products.
    3. In a customer sales force structure, the sales force is organized along customer or industry lines.
        i. Separate sales forces may be set up for different industries, for serving current customers versus finding new ones, and for major accounts versus regular accounts.
        ii. Organizing the sales force around customers can help a company to become more customer focused and build closer relationships with important customers.
    4. A complex sales force structure is often used when a company sells a wide variety of products to many types of customers over a broad geographic area.
        i. Salespeople can be specialized by customer and territory, by product and territory, by product and customers, or by territory, product, and customer.
        ii. No single structure is best for all companies and all situations.
c. Once the company has set its structure, it is ready to consider sales force size.
    1. Many companies use some form of workload approach to set sales force size.

<table>
<tr><td>i.</td><td>Using this approach, a company first groups accounts into different classes according to size, account status, or other factors that relate to the amount of effort required to maintain them.</td></tr>
<tr><td>ii.</td><td>The company then determines the number of salespeople needed to call on each class of accounts the desired number of times.</td></tr>
</table>

d. Sales management must also decide who will be involved in the selling effort and how various sales and sales support people will work together.

    1. The company may have an outside sales force, an inside sales force, or both.

        i. Outside salespeople travel to call on customers.

        ii. Inside salespeople conduct business from their offices via telephone or visits from prospective buyers.

        iii. Inside salespeople include support people, sales assistants, Web sellers, and telemarketers.

    2. Most companies are now using team selling to service large, complex accounts.

        i. Teams might include experts from any area or level of the selling firm, including sales, marketing, technical and support services, R&D, engineering, operations, finance, and others.

        ii. The move to team selling mirrors similar changes in customers' buying organizations.

        iii. Team selling does have some pitfalls. Selling teams can confuse or overwhelm customers who are used to working with only one salesperson. Salespeople who are used to having customers all to themselves may have trouble learning to work with and trust others on a team. Difficulties in evaluating individual contributions to the team selling effort can create some compensation issues.

Recruiting and Selecting Salespeople

e. At the heart of any successful sales force operation is the recruitment and selection of good salespeople.

f. According to the Gallup Management Consulting Group's research, the best salespeople possess four key talents: intrinsic motivation, disciplined work style, the ability to close sales, and the ability to build relationships with customers.

g. When recruiting, companies should analyze the sales job itself and the characteristics of its most successful salespeople to identify the traits needed by a successful salesperson in their industry.

h. Recruiting will attract many applicants from whom the company must select the best.

i. The selection process can vary from a single informal interview to lengthy testing and interviewing.

1. Many companies give formal tests to sales applicants.
2. Tests typically measure sales aptitude, analytical and organizational skills, personality traits, and other characteristics.

Training Salespeople

j. New salespeople may spend anywhere from a few weeks or months to a year or more in training.
k. The average initial training period is 4 months. Then, most companies provide continuing sales training via seminars, sales meetings, and the web throughout the salesperson's career.
l. Training programs have several goals.
    1. Salespeople need to know and identify with the company, so most training programs begin by describing the company's history and objectives, its organization, its financial structure and facilities, and its chief products and markets.
    2. Salespeople also need to know customers' and competitors' characteristics, so the training program teaches them about competitors' strategies.
    3. Many companies are adding Web-based training to their sales training programs. Such training ranges from simple text-based product information to Internet-based sales exercises to sophisticated simulations that re-create sales calls.

Compensating Salespeople

m. Compensation is made up of several elements: a fixed amount, a variable amount, expenses, and fringe benefits.
    1. The fixed amount, usually a salary, gives the salesperson some stable income.
    2. The variable amount, which might be commissions or bonuses based on sales performance, rewards the salesperson for greater effort.
    3. Expense allowances, which repay salespeople for job-related expenses, let salespeople undertake needed and desirable selling efforts.
    4. Fringe benefits, such as paid vacations, sick leave, accident benefits, pensions, and life insurance, enhance job satisfaction.
n. Management must decide what mix of these compensation elements makes the most sense for each sales job.
    1. Different combinations of fixed and variable compensation give rise to four basic types of compensation plans.
        i.   Straight salary.
        ii. Straight commission.
        iii. Salary plus bonus.
        iv. Salary plus commission.

2.   Compensation should direct the sales force toward activities that are consistent with overall marketing objectives. Table 13-1 shows an illustration of a compensation plan.

Supervising Salespeople

o.   Through supervision, the company directs and motivates the sales force to do a better job.

     1.   Companies vary in how closely they supervise their salespeople. They use various tools.

          i.   An annual call plan shows which customers and prospects to call on in which months and which activities to carry out.

          ii.   A time-and-duty analysis could be performed. Figure 13-2 shows how salespeople spend their time.

     2.   Many firms have adopted sales force automation systems, computerized sales force operations for more efficient order-entry transactions, improved customer service, and better salesperson decision-making support.

p.   Sales managers must also motivate salespeople.

     1.   Management can boost sales force morale and performance through its organizational climate, sales quotas, and positive incentives.

          i.   Organizational climate describes the feeling that salespeople have about their opportunities, value, and rewards for good performance.

          ii.   Many companies adopt sales quotas, which are standards stating the amount they should sell and how sales should be divided among the company's products. Compensation is often related to how well salespeople meet their quotas.

          iii.   Various positive incentives, such as sales meetings, sales contests, and honors, merchandise and cash awards, trips, and profit-sharing, are also used to motivate and reward sale force performance.

Evaluating Salespeople

q.   Management gets information about its salespeople in many ways.

     1.   Sales reports are weekly or monthly work plans and longer-term territory marketing plans.

     2.   Call reports are based on salespeople's completed activities.

     3.   Expense reports show what salespeople will be partly or wholly repaid.

r.   Formal evaluation forces management to develop and communicate clear standards for judging performance.

4. **The Personal Selling Process**

a. The selling process consists of several steps that the salesperson must master. These steps focus on the goal of getting new customers and obtaining orders from them.

Steps in the Selling Process

b. Figure 13-3 shows the selling process consisting of seven steps.

    1. The first step is prospecting, which is identifying qualified potential customers.

        i. Salespeople must often approach many prospects to get just a few sales.

        ii. Although the company supplies some leads, salespeople need skill in finding their own.

        iii. Salespeople also need to know how to qualify leads—identifying the good ones and screening out the poor ones.

        iv. Prospects can be qualified by looking at their financial ability, volume of business, special needs, location, and possibilities for growth.

    2. The preapproach step is where salespeople learn as much as possible about an organization and its buyers.

        i. Salespeople can consult standard industry and online sources, acquaintances, and others to learn about a company.

        ii. Salespeople should set call objectives, which may be to qualify a prospect, to gather information, or to make an immediate sale.

    3. The approach step occurs when salespeople meet and greet a buyer to get a relationship off to a good start.

        i. This step involves salespeople's appearance, opening lines, and the follow-up remarks.

    4. During the presentation step of the selling process, salespeople tell the product "story" to a buyer, presenting customer benefits and showing how the product solves the customer's problems.

        i. The need-satisfaction approach calls for good listening and problem-solving skills.

        ii. The qualities that buyers dislike most in salespeople include being pushy, late, deceitful, and unprepared or disorganized.

        iii. The qualities buyers value most include empathy, good listening, honesty, dependability, thoroughness, and follow-through.

    5. In handling objections, salespeople should use a positive approach, seek out hidden objections, ask the buyer to clarify any objections, take objections as opportunities to provide more information, and turn the objections into reasons for buying.

    6. Closing is the process of getting the order.

i.   Salespeople should know how to recognize closing signals from the buyer, including physical actions, comments, and questions.

ii.   Salespeople can use one of several closing techniques.

       a.   They can ask for the order.

       b.   They can review points of agreement.

       c.   They can offer to help write up the order.

       d.   They can ask whether the buyer wants this model or that one.

       e.   They can note that the buyer will lose out if the order is not placed now.

7.   The last step in the selling process is follow-up. This is necessary if salespeople want to ensure customer satisfaction and repeat business.

Personal Selling and Customer Relationship Management

c.   The principles of personal selling as just described are transaction-oriented; their aim is to help salespeople close a specific sale with a customer.

d.   In many cases, companies want profitable long-term relationships with customers they can win and keep.

e.   The sales force usually plays an important role in building and managing profitable customer relationships.

f.   Today's large customers favor suppliers who can sell and deliver a coordinated set of products and services to many locations and who can work closely with customer teams to improve products and processes.

5.   Direct Marketing

a.   With the trend toward more narrowly targeted or one-to-one marketing, many companies are adopting direct marketing, either as a primary marketing approach or as a supplement to other approaches.

b.   Direct marketing consists of direct connections with carefully targeted individual consumers to both obtain an immediate response and cultivate lasting customer relationships.

The New Direct-Marketing Model

c.   Most companies still use direct marketing as a supplementary channel or medium for marketing their goods.

d.   For many companies today, however, direct marketing is more than just a supplementary channel or medium.

     1.   Especially in its newest transformation—Internet marketing and e-commerce—direct marketing constitutes a new and complete model for doing business.

     2.   This new direct model is rapidly changing the way companies think about building relationships with customers.

Benefits and Growth of Direct Marketing

e.   For buyers, direct marketing is convenient, easy to use, and private.

    1.   Direct marketing gives buyers ready access to a wealth of products and information, both at home and around the globe.

    2.   Direct marketing is immediate and interactive—buyers can interact with sellers by phone or on the seller's website to create exactly the configuration of information, products, or services they desire, and then order them on the spot.

f.   For sellers, direct marketing is a powerful tool for building customer relationships.

    1.   Using database marketing, today's marketers can target small groups or individual consumers, tailor offers to individual needs, and promote these offers through personalized communications.

    2.   Direct marketing can be timed to reach prospects at just the right moment.

    3.   Direct marketing gives access to buyers that the company could not reach through other channels.

    4.   Direct marketing offers a low-cost, efficient alternative for reaching their markets.

g.   As a result of these advantages to both buyers and sellers, direct marketing has become the fastest-growing form of marketing.

Customer Databases and Direct Marketing

h.   A customer database is an organized collection of comprehensive data about individual customers or prospects, including geographic, demographic, psychographic, and behavioral data.

    1.   The database can be used to locate good potential customers, tailor products and services to the special needs of targeted consumers, and maintain long-term customer relationships.

    2.   A customer mailing list, in contrast, is simply a set of names, addresses, and telephone numbers.

i.   Companies use their databases in many ways.

    1.   They can use a database to identify prospects and generate sales leads by advertising products or offers.

    2.   They can mine their databases to learn about customers in detail, and fine-tune their market offerings and communications to the preferences and behaviors of target segments or individuals.

    3.   The company database can be an important tool for building stronger long-term customer relationships.

Forms of Direct Marketing

j.   The major forms of direct marketing are shown in Figure 13-4.

    1.   Telephone marketing uses the telephone to sell directly to consumers and business customers.

        i.   It has become the major direct-marketing communication tool.

ii.　B2B telephone marketing now accounts for more than 60% of all telephone marketing sales.

iii.　Marketers use outbound telephone marketing to sell directly to consumers and businesses.

iv.　Inbound toll-free 800 numbers are used to receive orders from television and print ads, direct mail, or catalogs.

v.　Properly designed and targeted telemarketing provides many benefits, including purchasing convenience and increased product and service information.

vi.　However, the recent explosion in unsolicited telephone marketing has annoyed many consumers.

　　a.　The FTC opened its "Do Not Call List" in mid-2003, and to date more that 87 million phone numbers have been registered.

2.　Direct-mail marketing involves sending an offer, announcement, reminder, or other item to a person at a particular address.

i.　Using highly selective mailing lists, direct marketers send out millions of mail pieces each year.

ii.　Direct mail has proved successful in promoting all kinds of products from books to gourmet foods to industrial products.

iii.　Direct mail is also heavily used by charities to raise billions of dollars each year.

iv.　The direct mail industry constantly seeks new methods and approaches. CDs and DVDs are among the fastest growing direct-mail media.

v.　Three new forms of mail delivery have become popular.

　　a.　Fax mail: Marketers now routinely send fax mail announcing special offers, sales, and other events to prospects and customers with fax machines.

　　b.　Email: Many marketers now send sales announcements, offers, product information, and other messages to email addresses. They may be resented as junk mail or SPAM if sent to people who have no interest in them. Smart marketers target their direct mail carefully.

　　c.　Voice mail: Some marketers have set up automated programs that exclusively target voice mailboxes and answering machines with prerecorded messages.

3.　Catalog marketing has grown explosively during the past 25 years.

i.　Annual catalog sales are expected to grow to more than $175 billion in 2008.

ii.　Web-based catalogs present a number of benefits.

　　a.　They save on production, printing, and mailing costs.

　　b.　They can offer an almost unlimited amount of merchandise.

             c. They allow real-time merchandising.

             d. Online catalogs can be spiced up with interactive entertainment and promotional features.

        iii. Web-catalogs present some challenges.

             a. They are passive and must be marketed.

             b. Attracting customers is much more difficult for a Web catalog than for a print catalog.

4.   Direct-response television marketing takes one of two major forms.

        i. Direct-response advertising is where direct marketers air television spots, often 60 or 120 seconds long, that persuasively describe a product and give customers a toll-free number for ordering.

        ii. Infomercials are 30-minute advertising programs for a single product.

             a. For years, infomercials have been associated with somewhat questionable pitches.

             b. But major corporations have been using infomercials to sell their wares.

        iii. With widespread distribution on cable and satellite television, the top three shopping networks combined now reach 248 million homes worldwide, selling more than $7.5 billion of goods each year.

5.   Kiosks are information and ordering machines in stores, airports, and other locations.

Integrated Direct Marketing

k.   Too often, a company's individual direct-marketing efforts are not well integrated with one another or with elements of its marketing and promotion mixes.

l.   Integrated direct marketing is a power approach that involves using coordinated multiple-media, multiple-stage campaigns.

Public Policy and Ethical Issues in Direct Marketing

m.   The aggressive and sometimes shady tactics of a few direct marketers can bother or harm consumers, giving the industry a black eye.

n.   Direct-marketing excesses sometimes annoy or offend consumers.

      1.   Dinner-time or late-night phone calls are especially bothersome.

      2.   So-called heat merchants design mailers and write copy intended to mislead buyers.

      3.   Some direct marketers pretend to be conducting research surveys when they are actually asking leading questions to screen or persuade customers.

o.   Invasion of privacy is perhaps the toughest public policy issue now confronting the direct-marketing industry.

1. It seems that almost every time consumers enter a sweepstakes, apply for a credit card, take out a magazine subscription, or order products by mail, telephone, or the Internet, their names are entered into some company's already bulging database.
2. Although consumers often benefit from database marketing, many critics worry that marketers may know too much about consumers' lives.

p. The direct marketing industry is addressing issues of ethics and public policy.
1. Direct marketers know that, left untended, such problems will lead to increasingly negative consumer attitudes, lower response rates, and calls for more restrictive state and federal legislation.
2. Most direct marketers want the same things that consumers want: honest and well-designed marketing offers targeted only toward consumers who will appreciate and respond to them.

Creative Marketing Exercises Designed to Reinforce the Concepts!!! (Suggested answers to these exercises can be found at the end of the Study Guide.)

1. Identify five alternative names for a salesperson.
2. Visit www.monster.com and view the list of sales job openings in your area. Identify a company that shares your philosophy on personal selling.
3. Contact the circulation desk at your local newspaper and inquire about how they find new customers. Explain how they divide their target markets.
4. You are the manager of a phone bank charged with selling new long distance phone services. Script out a sales dialogue for your employees to follow.
5. Your boss offers you a job selling furnaces for straight commission or dishwashers for a salary plus commission. Which would you choose and why?
6. Besides cash bonuses, what other kinds of incentives do you think would motivate employees?
7. Ask the salesperson at your favorite retail store how his/her performance is evaluated. Do sales goals play a role in determining whether or not he/she is successful?
8. You are selling encyclopedias door to door. Outline your sales technique using all the steps in the selling process.
9. Identify 5 companies that use direct marketing as an effective sales tool. What makes their use of direct sales more successful than others?
10. Videotape your favorite infomercial and be prepared to share why you think it is a great selling tool.

"Linking the Concepts" (#1) – Suggestions/Hints

1. Most people's perceptions of salespeople have changed over the years. No longer are they mainly men wearing polyester plaid pants, white belt, white shoes, chewing on a toothpick, and wearing that "I will rip you off" smile. Today's salespeople can be male or female, very professional, highly educated, and concerned about their customers' needs. They have college degrees, sometimes as technical as engineering and chemistry. They may have an MBA, or even a PhD. Today's salespeople will have access to technical sales tools to help them through the sales process. They are more concerned with relationship selling than making a quick buck. Today's selling profession is a professional career.

2. Using the steps in sales force management, CDW could go through the process as follows:

   Designing sales force strategy and structure: CDW owes its success to its highly effective "clicks and people" direct marketing strategy. CDW's direct model combines good old fashion high-touch personal selling with a modern high-tech Web presence to build lasting one-to-one customer relationships. This strategy is fueled by a genuine passion for solving customer problems. Under CDW's "Circle of Service" philosophy, "everything revolves around the customer."

   Recruiting and selecting salespeople: CDW needs to hire quality salespeople who possess the skills of teamwork, problem solving, and excellent communication. These skills will be needed to ensure quality customer relationships.

   Training salespeople: Before they make a single sales call, new account managers complete a six-week orientation and then a six-month training program. CDW University's College of Sales offers intensive schooling in the science behind the company's products and in the art of consultative selling. Tenured account managers receive ongoing training to enhance their relationship-selling skills.

   Compensating salespeople: CDW will need to find ways to compensate its sales force. Examples will include direct and indirect compensation packages.

   Supervising salespeople: Sales managers will need to motivate and supervise their sales force to ensure the continuous improvement of customer relationships.

   Evaluating salespeople: Sales managers will need to continuously and periodically evaluate their sales force to enhance and ensure quality customer service and the fulfillment of CDW's corporate goals.

3. The hiring process of most companies uses the same steps. Most companies recruit their sales forces from a number of different sources. Some of those sources are promoting from within, local colleges, newspaper and online want ads, walk-ins, personnel agencies, and "stealing" from competitors. The selection process is determined by company and customer needs. Prospects need to be self-disciplined, energetic, and have outstanding problem solving, communication, and intellectual skills. After the correct prospect is selected, then most companies need to train the new sales force. Training includes company, product, and competitor information. The new sales force will need to be taught the company's sales/ordering processes, and other company procedures/policies.

Compensation usually falls into either the salary, commission, or a combination of both categories. There are also indirect compensation packages that need to be considered, such as insurance, expense accounts, or a company car. Sales managers will motivate, train, and supervise their sales force with sales goals, cost goals, and periodic reviews.

"Linking the Concepts" (#2) – Suggestions/Hints

1. Some consumers buy into timeshares at resorts because of the direct marketing materials they receive. Consumers like to respond to these marketing tools because there is a simple, inexpensive exchange offered. That exchange may be a free weekend at some nice resort area for usually one hour of the consumer's time to listen to a sales presentation about the resort. There seems to be little risk involved. Most consumers see that they are getting a great weekend "free" or with little cost, other than time.

   A lot of direct marketing materials for "free" personal finance assistance are rejected by consumers. Most consumers feel that these are not legitimate organizations or products so they don't trust what they are reading in the marketing materials. If a company that is getting a lot of its direct marketing materials rejected by consumers wants to increase its chances of being accepted, it will need to address this fact. These companies will need to overcome the trust factor that they currently do not have. Consumers have seen too many news stories about people being ripped of by companies that used direct marketing materials to reel them in to their businesses.

2. Some examples of direct marketing materials that consumers can receive in an average week are listed below. Some are more popular than others (catalogs vs. credit card applications). Telemarketers seem to be the least preferred method of direct marketing by consumers. Most direct marketing materials are effective at hitting their target markets due to the increased use of technology and the study of consumer demographics and geographics.

Examples:
Catalogs
Bill inserts
Flyers on doors or mailboxes
Credit card applications
Postcards offering free weekends at resorts
Editorial material
Infomercials on TV
QVC, Home Shopping Network
Inserts in newspapers and magazines

Marketing Adventure Exercises (Suggested answers to these exercises can be found at the end of the Study Guide.)

(Visit www.prenhall.com/adventure for advertisements.)

1.   Financial   Ameritrade

Compare advertising and personal selling. Is the selected ad effective as both an advertisement and as a replacement for a personal sales force?

2.   Services   Student choice

Because personal selling can be the most expensive element in the promotion mix, a small company might use advertising in place of its own sales force. Which ad in the services category might be used in such circumstances?

3.   Financial   Student choice

What is direct marketing? Review the ads in the financial category and select one that you believe can be used effectively as a direct marketing piece.

4.   Student choice

In the text, direct marketing is referred to as both direct distribution and as an element of the marketing communications mix. Explain this and find an ad that shows a product that approaches direct marketing in both these contexts.

5.   Student choice

Direct marketing is beneficial to both buyers and sellers. Discuss the benefits to both, and then select an ad that highlights the benefits to buyers.

6. Student choice

List and describe each of the main forms of direct marketing. Select an ad and alter it so it may be used in direct marketing.

7. Financial   Student choice

Telephone marketing directly to consumers and business customers has become a major direct-marketing communication tool, accounting for nearly 40% of all direct-marketing media expenditures. Which ad in the financial category can be related to telephone marketing?

8. Nonprofit   Student choice

What are the advantages of direct mail marketing? Select an ad from the nonprofit category that would function as a direct mail piece.

SUM IT UP!!!!!!

Using only this page, sum up all of the concepts and terms discussed in Chapter 13 – "Communicating Customer Value: Personal Selling and Direct Marketing". Here is your chance to make sure you know and understand the concepts!!!!

Chapter 14
Marketing in the Digital Age

Previewing the Concepts: Chapter Objectives

1. Discuss how the digital age is affecting both consumers and the marketers who serve them.
2. Explain how companies have responded to the Internet and other powerful new technologies with e-business strategies, and how these strategies have resulted in benefits to both buyers and sellers.
3. Describe the four major e-commerce domains.
4. Discuss how companies go about conducting e-marketing to profitably deliver more value to customers.
5. Overview the promise and challenges that e-commerce presents for the future.

JUST THE BASICS

Chapter Overview

Some say the new digital technologies have created a new economy. Few would disagree that the Internet and other powerful new connecting technologies are having a dramatic impact on marketers and buyers. Companies need to retain most of the skills and practices that have worked in the past, but they will need to add major new competencies and practices if they hope to grow and prosper in the changing digital environment.

Much of the world's business today is carried out over networks that connect people and companies. Intranets, extranets, and the Internet itself have all changed the way companies do business, and customers find the products and services they want. The explosive worldwide growth in Internet usage forms the heart of the so-called new economy. But the Internet has also allowed new companies to compete, and the formation of new types of intermediaries and new forms of channel relationships caused existing firms to re-examine how they served their markets. Finally, e-commerce continues to offer both great promise and many challenges for the future.

Conducting business in the new digital age will call for a new model for marketing strategy and practice. Some strategists envision a day when all buying and selling will involve direct electronic connections between companies and their customers. But the fact is that today's marketing requires a mixture of old economy and new economy thinking and action.

E-business involves the use of electronic platforms such as intranets, extranets, and the Internet to conduct a company's business. E-commerce is more specific than e-business.

E-commerce involved buying and selling processes supported by electronic means, primarily the Internet. E-marketing is the marketing side of e-commerce.

There are several e-marketing domains, including business-to-consumer (B2C), business-to-business (B2B), consumer-to-consumer (C2C), and consumer-to-business (C2B). Each of these domains meets specific needs of each of the segments addressed, and they all continue to grow.

E-commerce is conducted in many ways. Companies can be "click-only" in that they are located only on the Internet. They include e-tailers, search engines and portals, Internet Service Providers (ISPs), transaction sites, content sites, and enabler sites. Many companies today, however, are "click-and-mortar" companies, because they maintain their traditional channels of distribution while simultaneously providing an Internet channel.

There are various types of websites. Corporate websites typically offer a rich variety of information and other features in an effort to answer customer questions, build closer customer relationships, and generate excitement about the company. Marketing websites engage consumers in an interaction that will move them closer to a direct purchase or other marketing outcome. Online advertising includes such things as banner ads and tickers; skyscrapers, which are tall, skinny ads at the side of a web page; rectangles; and interstitials—online ads that pop up between changes on a website. Viral marketing involves creating an email message or other marketing event that is so infectious that customers will want to pass it along to their friends.

E-commerce continues to offer both great promise and many challenges for the future. Online marketing will become a successful model for some companies. However, there is a darker side to Internet marketing. One major concern is profitability, especially for B2C dot-coms. Although expanding rapidly, online marketing still reaches only a limited marketplace.

There are also broader ethical and legal questions. Online privacy is perhaps the number one e-commerce concern. Many consumers also worry about online security, as well as the privacy rights of children. Many companies have responded to consumer privacy and security concerns with actions of their own. Still, examples of companies aggressively protecting their customers' personal information are too few and far between.

Chapter Outline

1. **Introduction**
   a.   Chances are, when you think of shopping on the Web, you think of Amazon.com. Amazon.com first opened its virtual doors in mid-July 1995, selling books out of founder Jeff Bezos's garage in suburban Seattle.
   b.   It still sells books—by the millions. But it now sells products in a dozen other categories as well.

c.   In perfecting the art of online selling, it has also rewritten the rules of marketing. Its most ardent fans view Amazon.com as the model for businesses in the new digital age.

d.   But not everything has clicked smoothly for Amazon.com. It has more than 39 million customers in more than 220 countries, but Amazon.com didn't turn its first full-year profit until just last year, and those profits were modest.

e.   Amazon.com obsesses over making each customer's experience uniquely personal. Visitors to Amazon.com's Web site receive a unique blend of benefits: huge selection, good value, convenience, and what Amazon vice president Jason Kilar calls "discovery."

f.   In fact, Amazon.com has become so good at managing online relationships that many traditional "brick-and-mortar" retailers are turning to Amazon for help in adding more "clicks" to their "bricks."

g.   Recent technological advances have created a new digital age. Widespread use of the Internet and other powerful new technologies are having a dramatic impact on marketers and buyers.

2.   **The Digital Age**

a.   Much of the world's business today is carried out over digital networks that connect people and companies.

  1.   Intranets are networks that connect people within a company to each other and to the company network.

  2.   Extranets connect a company with its suppliers, distributors, and other outside partners.

  3.   The Internet is a vast public web of computer networks; it connects users of all types all around the world to each other and to an amazingly large "information repository."

b.   With the creation of the World Wide Web and web browsers in the 1990s, the Internet was transformed from a mere communication tool into a certifiably revolutionary technology.

c.   The explosive worldwide growth in Internet usage forms the heart of the so-called new economy. The Internet has been the revolutionary technology of the new millennium, empowering consumers and businesses alike with blessings of connectivity.

d.   In 2004, Internet penetration in the U.S. had reached 68 percent, with more than 202 million people not using the Internet. Some 14.6 percent of the world population---more than 938 million worldwide—are now online.

e.   The average U.S. Internet user spends 28 hours a month surfing the Web at home, plus another 76 hours a month at work.

f.   The Internet and other digital technologies have given marketers a whole new way to reach and serve customers.

  1.   The amazing success of early click-only companies such as Amazon.com caused brick and mortar manufacturers and retailers to reexamine how they served their markets.

2. Now, almost all of these traditional companies have become click-and-mortar competitors.

g. It's hard to find a company today that doesn't have a substantial Web presence.

3. Marketing Strategy in the Digital Age

a. Conducting business in the new digital age will call for a new model for marketing strategy and practice.

b. The Internet is revolutionizing how companies create value for customers and build customer relationships.

c. The digital age has fundamentally changed customers' notions of convenience, speed, price, product information, and service.

d. The fact is that today's marketing requires new thinking and action.

1. Companies need to retain most of the skills and practices that have worked in the past.

2. But they will also need to add major new competencies and practices if they hope to grow and prosper in the new environment.

E-Business, E-Commerce, and E-Marketing in the Digital Age

e. E-business involves the use of electronic platforms—intranets, extranets, and the Internet—to conduct a company's business.

1. Almost every company has set up a website to inform about and promote their products and services.

2. Most companies have also created intranets to help employees communicate with each other and to access information found in the company's computers.

3. Companies also set up extranets with their major suppliers and distributors to enable information exchange, orders, transactions, and payments.

f. E-commerce is more specific than e-business.

1. E-business includes all electronics-based information exchanges within or between companies and customers.

2. In contrast, e-commerce involves buying and selling processes supported by electronic means, primarily the Internet.

3. E-markets are "marketspaces" rather than physical marketplaces.

i. Sellers use e-markets to offer their products and services online.

ii. Buyers use them to search for information, identify what they want, and place orders using credit or other means of electronic payment.

4. E-commerce includes e-marketing and e-purchasing.

i. E-marketing is the marketing side of e-commerce. It consists of company efforts to communicate about, promote, and sell products and services over the Internet.

199

ii.   E-purchasing is the flip side of e-marketing. It is the buying side of e-commerce. It consists of companies purchasing goods, services, and information from online suppliers.

Benefits to Buyers

g.   Internet buying benefits both final buyers and business buyers in many ways.
1.   It can be convenient.
2.   Buying is easy and private.
3.   The Internet often provides buyers with greater product access and selection.
4.   E-commerce channels also give buyers access to a wealth of comparative information about companies, products, and competitors.
5.   Online buying is interactive and immediate.

Benefits to Sellers

h.   There are also many benefits to sellers.
1.   The Internet is a powerful tool for customer relationship building.
2.   The Internet and other electronic channels can also reduce costs and increase speed and efficiency.
3.   E-marketing can also offer greater flexibility, allowing the marketer to make ongoing adjustments to its offers and programs.
4.   The Internet is a truly global medium that allows buyers and sellers to click from one country to another in seconds.

4.   **E-Marketing Domains**
a.   The four major e-marketing domains are shown in Figure 14-1.

B2C (Business-to-Consumer)

b.   The popular press has paid the most attention to B2C (business-to-consumer) e-commerce—the online selling of goods and services to final consumers.
c.   Online consumer buying continues to grow at a healthy rate.
d.   Last year, U.S. consumers spend $117 billion online, and consumer spending online is expected to exceed $316 billion by 2010, accounting for 12 percent of total sales.
1.   As the Web has matured, Internet demographics have changed significantly.
2.   Whereas 66 percent of adults use the Internet, 87 percent of teens go online. Consumers aged 50 and older make up almost 20 percent of the online population. More than 22 million Americans over 65 are expected to be online by 2009.
3.   E-marketing targets people who actively select which websites they will visit and what marketing information they will receive about which products and under what conditions.

e. Consumers can find a Web site for buying almost anything. The Internet is most useful for products and services when the shopper seeks greater ordering convenience or lower costs.

B2B (Business-to-Business)

f. Consumer goods sales via the web are dwarfed by B2B (business-to-business) e-commerce.

g. Most major business-to-business marketers now offer product information, customer purchasing, and customer support services online.

h.

i. Increasingly, online sellers are setting up their own private trade exchanges. These exchanges link a particular seller with its trading partners.
    1. Private exchanges give sellers greater control over product presentation and allow them to build deeper relationships with buyers and sellers by providing value-added services.

C2C (Consumer-to-Consumer)

j. Much C2C (consumer-to-consumer) e-commerce and communication occurs on the web between interested parties over a wide range of products and subjects.
    1. C2C involves interchanges of information through Internet forums that appeal to specific special-interest groups.
    2. Blogs are online journals where people post their thoughts, usually on a narrowly defined topic. Blogs can be about anything, from politics to baseball to haiku or car repair.
    3. Many marketers are tapping into blogs as a medium for reaching carefully targeted consumers. Blogs offer a fresh, original, personal, and cheap way to reach today's fragmented audiences.

k. C2C means that online visitors don't just view consumer product information. Increasingly, they create it. They join Internet interest groups to share information, with the result that "word of web" is joining "word of mouth" as an important buying influence.

C2B (Consumer-to-Business)

l. C2B (consumer-to-business) e-commerce allows today's consumers to communicate more easily with companies.

m. Most companies now invite prospects and customers to send in suggestions and questions via company websites.

5. **Marketing on the Web**
    a. Companies of all types are now marketing online. The different types of e-marketers are shown in Figure 14-2.

201

<u>Click-Only versus Click-and-Mortar E-Marketers</u>

b. The Internet gave birth to a new species of e-marketers—the click-only dot-coms—which operate only online without any brick-and-mortar market presence.

c. Brick-and-mortar companies have now added e-marketing operations, transforming themselves into click-and-mortar competitors.

d. Click-only companies come in many shapes and sizes.

    1. E-tailers are dot-coms that sell products and services directly to final buyers via the Internet.

        i. This group includes search engines and portals.

        ii. Internet service providers (ISPs) are click-only companies that provide Internet and email connections for a fee.

        iii. Transaction sites take commissions for transactions conducted on their sites.

        iv. Content sites provide financial, research, and other information.

    2. In the late 1990s dot-coms reached astronomical levels but then collapsed in the year 2000 because of lavish spending and reliance on spin and hype.

    3. Now on firmer footing, dot-coms are surviving and even prospering in today's marketspace.

e. Many established companies moved quickly to open websites providing information about their companies and products.

    1. Most resisted adding e-commerce to their sites. They worried that this would produce channel conflict—that selling their products or services online would be competing with their offline retailers and agents.

    2. However, they soon realized that the risks of losing business to online competitors were even greater than the risks of angering channel partners.

    3. Most established brick-and-mortar companies are now prospering as click-and-mortar companies.

    4. Most of these click-and-mortar companies have found ways to resolve channel conflicts.

    5. Established companies have known and trusted brand names and greater financial resources. They have large customer bases, deeper industry knowledge and experience, and good relationships with key suppliers. By combining online marketing and established brick-and-mortar operations, they can offer customers more options.

<u>Setting Up an Online Marketing Presence</u>

f. Companies can conduct e-marketing in any of the four ways shown in Figure 14-3.

g. The first step in conducting e-marketing is to create a website.
   1. The most basic type is a corporate website.
      i. These sites are designed to build customer goodwill and to supplement other sales channels, rather than to sell the company's products directly.
      ii. Corporate websites typically offer a rich variety of information and other features in an effort to answer customer questions, build closer customer relationships, and generate excitement about the company.
      iii. These sites generally provide information about the company's history, its mission and philosophy, and the products and services it offers.
   2. Other companies create a marketing website.
      i. These sites engage consumers in an interaction that will move them closer to a direct purchase or other marketing outcome.
      ii. Such sites might include a catalog, shopping tips, and promotional features such as coupons, sales events, or contests.
h. Creating a website is one thing; getting people to visit the site is another.
   1. The key is to create enough value and excitement to get consumers to come to the site, stick around, and come back again.
   2. This means that companies must constantly update their sites to keep them current, fresh, and exciting.
   3. A key challenge is designing a website that is attractive on first view and interesting enough to encourage repeat visits.
   4. To attract new visitors and to encourage revisits, one expert suggests playing close attention to the seven Cs of effective website design.
      i. Context—the site's layout and design.
      ii. Content—the text, pictures, sound, and video that the website contains.
      iii. Community—the ways that the site enables user-to-user communication.
      iv. Customization—the site's ability to tailor itself to different users or to allow users to personalize the site.
      v. Communication—the ways the site enables site-to-user, user-to-site, or two-way communication.
      vi. Connection—the degree that the site is linked to other sites.
      vii. Commerce—the site's capabilities to enable commercial transactions.
   5. Ultimately, it is the value of the site's content that will attract visitors, get them to stay longer, and bring them back for more.
i. E-marketers can use online advertising to build their Internet brands or to attract visitors to their websites.

1. Online ads that pop up while Internet users are surfing online include banner ads and tickers.
2. Skyscrapers are tall, skinny ads at the side of a web page, while rectangles are boxes that are much larger than a banner.
3. Interstitials are online ads that pop up between changes on a website.
4. Rich media ads incorporate animation, video, sound, and interactivity.
5. Content sponsorships are another form of Internet promotion. Many companies gain name exposure on the Internet by sponsoring special content on various websites, such as news or financial information.
6. Internet companies can develop alliances and affiliate programs, in which they work with other companies online and offline to promote each other.
7. Online marketers use viral marketing, the Internet version of word-of-mouth marketing. Viral marketing involves creating an email message or other marketing event that is so infectious that customers will want to pass it along to their friends.
8. Online advertising serves a useful purpose, especially as a supplement to other marketing efforts.
   i. Online advertising should be as much as 10 to 15 percent of the overall media mix in low-involvement products such as packaged goods.
   ii. Online advertising opens a two-way exchange to better educate consumers about products.
j. The popularity of blogs and other Web forums has resulted in a rash of commercially sponsored websites called web communities, which take advantage of the C2C properties of the Internet.
   1. Such sites allow members to congregate online and exchange views on issues of common interest.
   2. Visitors to these Internet neighborhoods develop a strong sense of community.
      i. Such communities are attractive to advertisers because they draw consumers with common interests and well-defined demographics.
      ii. Cyberhood consumers visit frequently and stay online longer, increasing the chance of meaningful exposure to the advertiser's message.
      iii. Web communities can be either social or work related.
k. Email has exploded onto the scene as an important e-marketing tool.
   1. To compete effectively in this ever-more-cluttered email environment, marketers are designing "enriched" email messages that are animated, interactive, and personalized messages full of streaming audio and video.

2. The recent explosion of spam—unsolicited, unwanted commercial email messages that clog up your emailboxes—has produced consumer frustration and anger.

6. The Promise and Challenges of E-Commerce

a. E-commerce continues to offer both great promise and many challenges for the future.

The Continuing Promise of E-Commerce

b. Its most ardent apostles still envision a time when the Internet and e-commerce will replace magazines, newspapers, and even stores as sources of information and buying.

c. Online marketing will become a successful business model for some companies. However, for most companies, online marketing will remain just one important approach to the marketplace that works alongside other approaches in a fully integrated marketing mix.

d. Eventually, as companies become more adept at integrating e-commerce with their everyday strategy and tactics, the "e" will fall away from e-business or e-marketing.

The Web's Darker Side

e. Along with its considerable promise, there is a "darker side" to Internet marketing.

f. One major concern is profitability, especially for B2C dot-coms.
   1. Surprisingly few Internet companies are profitable.
   2. One problem is that, although expanding rapidly, online marketing still reaches only a limited marketspace.
   3. The web audience is becoming more mainstream, but online users still tend to be somewhat more upscale and better educated than the general population.
   4. Another problem is that the Internet offers millions of Web sites and a staggering volume of information.
   5. A great number of click-only online retailers are small, niche marketers, and niches can become crowded and competitive.

g. From a broader societal viewpoint, Internet marketing practices have raised a number of ethical and legal questions.
   1. Online privacy is perhaps the number one e-commerce concern.
      i. Most online marketers have become skilled at collecting and analyzing detailed consumer information.
      ii. This may leave consumers open to information abuse if companies make unauthorized use of the information in marketing their products or exchanging databases with other companies.
   2. Many consumers worry about online security.

        i.   Consumers fear that unscrupulous snoopers will eavesdrop on their online transactions or intercept their credit card numbers and make unauthorized purchases.

        ii.   Companies doing business online fear that others will use the Internet to invade their computer systems for the purposes of commercial espionage or even sabotage.

        iii.   Of special concern are the privacy rights of children.

        iv.   Many companies have responded to consumer privacy and security concerns with actions of their own.

        v.   Still, examples of companies aggressively protecting their customers' personal information are too few and far between.

3. Consumers are also concerned about Internet fraud, including identity theft, investment fraud, and financial scams.

4. There are also concerns about segmentation and discrimination on the Internet.

        i.   Some social critics and policy makers worry about the so-called digital divide—the gap between those who have access to the latest Internet and information technologies and those who don't.

        ii.   A final Internet marketing concern is that of access by vulnerable or unauthorized groups.

h. As it continues to grow, online marketing will prove to be a powerful tool for building customer relationships, improving sales, communicating company and product information, and delivering products and services more efficiently and effectively.

Creative Marketing Exercises Designed to Reinforce the Concepts!!! (Suggested answers to these exercises can be found at the end of the Study Guide.)

1. Visit the web site for the United States Postal Service and discuss the significance of offering online services to customers.

2. Visit www.qvc.com and compare the use of the "marketspace" with the use of the television "marketplace."

3. Identify 5 examples of online companies that have enjoyed a quick rise to success. Outline why you think these companies have been successful where others have not.

4. Calculate the cost of running an interactive website for a year. Is this a cost effective method of doing business? Why or why not?

5. Locate 5 restaurants that use the Internet to bolster their business. What techniques do they use that make their sites unique?

6. Determine whether eBay would be considered to have a "Business to Business" marketing strategy or a "Consumer to Consumer" strategy. Justify your answer.

7. Locate 5 websites that fit each of the following criteria: college football, women's apparel, beer, news, and golf.

8. Visit www.gap.com and discuss the benefits of shopping online versus visiting the actual site.

9. Make a list of the "pop-up" ads you encounter in an hour of Internet surfing. Are these ads a selling tool or a nuisance?

10. Visit www.nyse.com and research the stock value of AOL Time Warner. How do you account for the variances over the past few years?

"Linking the Concepts" – Suggestions/Hints

1. Take, for example, the re-release of National Lampoon's *Animal House* on DVD. A search at Amazon.com finds that they have the DVD and it can be purchased from them. Amazon.com promises 24-hour delivery on that item. The search also finds anything that is related to the movie. A consumer can purchase T-shirts, books, and other movies related to *Animal House*.

2. If a consumer looks for the same DVD at Barnes and Noble's website, they will get the movie and they can purchase it from the site, as well. Barnes and Noble does offer the same 24-hour delivery on that item and the DVD is $0.01 cheaper. However, when the search is conducted at Barnes and Noble, the only items found are just the DVD. There are no other items related to the movie mentioned.

3. Amazon.com has more related products to offer.

    Local stores will have a hard time competing with these websites unless they can offer the consumer 24-hour delivery, too. That seems to be an important marketing tool on the websites. Local stores will not be able to carry the inventory that Amazon.com and Barnes and Noble carry. Both of those companies can extend their inventory offering through the inventories of their suppliers. That way they can offer the 24-hour delivery. This will be the challenge for local bookstore operators.

<u>Marketing Adventure Exercises (Suggested answers to these exercises can be found at the end of the Study Guide.)</u>

(Visit www.prenhall.com/adventure for advertisements.)

1.   Apparel   Johnsons
        General   Target2
        Internet   Exterminating

Your text states that "The formation of new types of intermediaries and new forms of channel relationships caused existing firms to reexamine how they served their markets." Review the selected ads, deciding whether they are strictly a "brick and mortar" company, a "click only" firm, or a hybrid of the two.

2.   Financial   Alliance, Cimb, Santander

Discuss the differences between customization and customerization.   Relate your discussion to the selected ads.

3.   Internet   Student choice

In an earlier chapter you learned that the newest transformation for direct marketing is Internet-based marketing.   Having learned about Internet marketing in this chapter, decide which ad in the Internet category you think is effective Internet-based marketing?

4.   Electronics   Student choice

Review the ads in the Electronics category.   Which would have a web site designed to sell the product, and which would have a web site that would strictly inform and connect the product with the customer?

5.   Financial   Regent

The selected ad encourages buyers to visit the web site and purchase the product.   What are the benefits of Internet buying?

6.   Internet   Shopnow

E-commerce provides sellers with many benefits.   Review the selected ad from the Internet category that represents an "e-seller," and discuss the benefits that seller experiences by having an Internet presence.

7.   Travel   Imperial, Imperial2

Review the two selected ads.   Distinguish between the B2B and B2C marketplace.

8.   Auto, Electronics   Student choice

List and explain the seven C's of effective web site design. Visit the website of any company with an ad in the Auto or Electronics category. Does their web site meet the effective web site criteria?

9.   Student choice

Some products are better suited for online marketing than others. Select 3 ads for products that are well suited for online marketing and select 3 ads for products that are less suitable for online marketing. Discuss your rationale.

SUM IT UP!!!!!!

Using only this page, sum up all of the concepts and terms discussed in Chapter 14 – "Marketing in the Digital Age". Here is your chance to make sure you know and understand the concepts!!!!

Chapter 15
The Global Marketplace

Previewing the Concepts: Chapter Objectives

1.   Discuss how the international trade system, economic, political-legal, and cultural environments affect a company's international marketing decisions.
2.   Describe three key approaches to entering international markets.
3.   Explain how companies adapt their marketing mixes for international markets.
4.   Identify the three major forms of international marketing organization.

JUST THE BASICS

Chapter Overview

This chapter looks at the special considerations that companies face when they market their brands globally. Advances in communication, transportation, and other technologies have made the world a much smaller place. Almost every firm, large or small, faces international marketing issues.

International trade is booming. Since 1969, the number of multinational corporations in the world's 14 richest countries has more than tripled, from 7,000 to 24,000. Global competition is intensifying. Foreign firms are expanding aggressively into new international markets, and home markets are no longer as rich in opportunity. If companies delay taking steps toward internationalizing, they risk being shut out of growing markets.

Before deciding to operate internationally, a company must thoroughly understand the international marketing environment. There are many issues to understand regarding the international trade system, such as tariffs, quotas, embargoes, and other barriers to entry. The World Trade Organization was established in the latest round of GATT negotiations. The WTO acts as an umbrella organization, overseeing GATT, the General Agreement on Trade in Services, and a similar agreement government intellectual property. In addition, the WTO mediates global disputes and imposes trade sanctions.

Two factors reflect a country's attractiveness as a market—the country's industrial structure and its income distribution. There are also several factors to contend with in the political-legal environment, as well as the cultural environment.

There are several factors that draw a company into the international arena. Global competitors might attack the company's domestic market by offering better products or lower prices. The company might want to counterattack these competitors in their home markets to tie up their resources. Or the company might discover foreign markets that present higher profit opportunities than the domestic market does. Before going abroad, the company must weigh several risks and answer many questions about its ability to operate globally.

Before going abroad, the company should try to define its international marketing objectives and policies. The company needs to choose how many countries it wants to market in, and it needs to decide on the types of countries to enter. Once the decision has been made to sell in a foreign country, the company must determine the best mode of entry.

Most companies start with exporting, using either indirect or direct exporting. Or it could go into a joint venture, through such means as licensing, contract manufacturing, management contracting, or joint ownership. Finally, the company could make a direct investment by developing assembly or manufacturing facilities. Each form of entry carries its own risks and rewards; the company must weigh these carefully before making final decisions.

Companies that operate in one or more foreign markets must decide how much, if at all, to adapt their marketing mixes to local conditions. At one extreme are global companies that use a standardized marketing mix, selling largely the same products and using the same marketing approaches worldwide. At the other extreme is an adapted marketing mix. In this case, the producer adjusts the marketing mix elements to each target market, bearing more costs but hoping for a larger market share and return. However, global standardization is not an all-or-nothing proposition. Rather, it is a matter of degree.

Finally, companies can manage their international marketing activities in at least three different ways: Most companies first organize an export department, then create an international division, and finally become a global organization.

Chapter Outline

1. **Introduction**
   a. Coca-Cola is a brand that is as American as baseball and apple pie. But from the beginning, Coke was destined to be more than just America's soft drink.
   b. First introduced in 1893, by 1900 Coca-Cola had already ventured beyond America's borders into numerous countries.
   c. Coca-Cola's worldwide success results from a skillful balancing of global standardization and brand building with local adaptation. For years, the company adhered to the mantra "Think globally, act locally." The company carefully adapts its mix of brands and flavors, promotions, price, and distribution to local customs and preferences in each market.
   d. As a result of its international marketing prowess, Coca-Cola dominates the global soft drink market.

2. Global Marketing in the Twenty-first Century

   a.   The world is shrinking rapidly with the advent of faster communication, transportation, and financial flows. Products developed in one country are finding enthusiastic acceptance in other countries.

   b.   International trade is booming.

       1.   Since 1969, the number of multinational corporations in the world has grown from 7,000 to more than 63,000.

       2.   Since 2003, world trade has been growing at 5 to 10 percent annually. World trade of products and services was valued at over 10.9 trillion dollars in 2004, which accounted for 20 percent of gross domestic product worldwide.

       3.   This trade growth is most visible in developing countries, which saw their share in world merchandise trade rise sharply to 31 percent, the highest since 1950.

   c.   If companies delay taking steps toward internationalizing, they risk being shut out of growing markets.

   d.   Although the need for companies to go abroad is greater today than in the past, so are the risks.

       1.   Companies that go global may face highly unstable governments and currencies, restrictive government policies and regulations, and high trade barriers.

       2.   Corruption is an increasing problem—officials in several countries often award business not to the best bidder but to the highest briber.

   e.   A global firm is one that, by operating in more than one country, gains marketing, production, R&D, and financial advantages that are not available to purely domestic competitors.

       1.   The global company sees the world as one market. It minimizes the importance of national boundaries and develops "transnational" brands.

       2.   The rapid move toward globalization means that all companies will have to answer some basic questions as to market position, competitors, strategies, resources, production, and strategic alliances.

3. Looking at the Global Marketing Environment

   a.   Before deciding whether to operate internationally, a company must thoroughly understand the international marketing environment.

<u>The International Trade System</u>

   b.   When selling to another country, the U.S. firm faces various trade restrictions.

       1.   A tariff is a tax levied by a foreign government against certain imported products. It may be designed either to raise revenue or to protect domestic firms.

2. A quota sets limits on the amount of goods the importing country will accept in certain product categories. The purpose of the quota is to conserve on foreign exchange and to protect local industry and employment.

3. Exchange controls limit the amount of foreign exchange and the exchange rate against other currencies.

4. Nontariff trade barriers are such things as biases against U.S. company bids or restrictive product standards or other rules that go against American product features.

c. The General Agreement on Tariffs and Trade (GATT) is a 58-year-old treaty designed to promote world trade by reducing tariffs and other international trade barriers.

1. Since its inception in 1948, there have been eight rounds of GATT negotiations to reassess trade barriers and set new rules for international trade.

   i. The first seven rounds reduced the average worldwide tariffs on manufactured goods from 45% to just 5%.

   ii. The Uruguay round, the most recent, lasted seven years.

      a. It reduced the world's remaining merchandise tariffs by 30%, boosting global merchandise trade by as much as 10%.

      b. It extended GATT to cover trade in agriculture and a wide range of services.

      c. It toughened international protection of copyrights, patents, trademarks, and other intellectual property.

      d. It established the World Trade Organization (WTO) to enforce GATT rules. The WTO mediates global disputes and imposes trade sanctions.

d. Certain countries have formed free trade zones or economic communities, which are groups of nations organized to work toward common goals in the regulation of international trade.

1. The European Community was formed in 1957.

   i. European unification offers tremendous trade opportunities for U.S. and other non-European firms.

   ii. It also poses threats.

      a. European companies will grow bigger and more competitive.

      b. Lower trade barriers inside Europe will create only thicker outside walls.

      c. Widespread adoption of the euro will decrease currency risk, making member countries with previously weak currencies more attractive markets.

2. In January 1994 the North American Free Trade Agreement (NAFTA) established a free trade zone among the United States, Mexico, and Canada.

3.   Given the apparent success of NAFTA, talks are underway to investigate establishing a Free Trade Area of the Americas (FTAA). This free trade zone would include 34 countries stretching form the Bering Strait to Cape Horn, with a population of 800 million.

4.   Other free trade areas have formed in Latin America and South America.

e.   The trend toward free trade zones has raised some concerns.

1.   In the United States, unions fear that NAFTA will lead to the further exodus of manufacturing jobs to Mexico, where wage rates are much lower.

2.   Environmentalists worry that companies that are unwilling to play by the strict rules of the U.S. Environmental Agency will relocate to Mexico, where pollution regulation has been lax.

Economic Environment

f.   The international marketer must study each country's economy.

g.   Two economic factors reflect the country's attractiveness as a market: the country's industrial structure and its income distribution.

1.   The country's industrial structure shapes its product and service needs, income levels, and employment levels.

i.   Subsistence economies are those where the vast majority of people engage in simple agriculture.

ii.   Raw material exporting economies are countries that are rich in one or more natural resources, but poor in other ways.

iii.   Industrializing economies are those where manufacturing accounts 10% to 20% of the country's economy.

iv.   Industrialized economies are major exporters of manufactured goods and investment funds.

2.   The second economic factor is the country's income distribution.

i.   Countries with subsistence economies may consist mostly of households with very low family incomes.

ii.   Industrialized nations may have low-, medium-, and high-income households.

iii.   Other countries may have households with only either very low or very high incomes.

iv.   Even low-income and developing economies may be attractive markets for all kinds of goods and luxuries.

Political-Legal Environment

h.   Nations differ greatly in their political-legal environments.

i.   At least four political-legal factors should be considered in deciding whether to do business in a given country.

1. In their attitudes toward international buying, some nations are quite receptive to foreign firms, and others are less accommodating.
2. Some countries such as India bother foreign business with import quotas, currency restrictions, and other limitations that make operating there a challenge.
3. Political stability is another issue.
4. Monetary regulations need to be studied.
j. Nations with too little hard currency may want to pay with other items instead of cash, which has led to the growing practice called countertrade. It takes several forms.
1. Barter involves the direct exchange of goods or services.
2. Compensation, or buyback, occurs when the seller sells a plant, equipment, or technology to another country and agrees to take payment in the resulting products.
3. Counterpurchase occurs when the seller receives full payment in cash but agrees to spend some portion of the money in the other country within a stated time period.

Cultural Environment
k. Each country has its own folk-ways, norms, and taboos. When designing global strategies, companies must understand how culture affects consumer reactions in each of its world markets. In turn, they must also understand how their strategies affect local cultures.
1. The seller must examine the ways consumers in different countries think about and use certain products before planning a marketing program.
2. Business norms and behaviors vary from country to country.
l. Some critics argue that "globalization" really means "Americanization."
1. These critics contend that exposure to American values and products erode other cultures and westernize the world.

4. **Deciding Whether to Go International**
a. Any of several factors might draw a company into the international arena.
1. Global competitors might attack the company's domestic market by offering better products or lower prices.
2. The company might want to counterattack those competitors in their home markets to tie up their resources.
3. The company might discover foreign markets that present higher profit opportunities than the domestic market does.
4. The company's home market might be stagnant or shrinking.
5. The company's customers might be expanding abroad and require international servicing.
b. Before going abroad, the company must weigh several risks and answer many questions about its ability to operate globally.

c. Because of the difficulties of entering international markets, most companies do not act until some situation or event thrust them into the global arena.

5. **Deciding Which Markets to Enter**
   a. Before going abroad, the company should try to define its international marketing objectives and policies.
   b. The company also needs to choose how many countries it wants to enter.
      1. Companies must be careful not to spread themselves too thin or to expand beyond their capabilities in too many countries too soon.
   c. The company needs to decide on the type of countries to enter.
      1. A country's attractiveness depends on the product, geographic factors, income and population, political climate, and other factors.
   d. After listing possible international markets, the company must screen and rank each one.
      1. Possible global markets should be ranked on several factors, including market size, market growth, cost of doing business, competitive advantage, and risk level.
      2. The goal is to determine the potential of each market, using indicators such as those shown in Table 15-1.

6. **Deciding How to Enter the Market**
   a. Once a company has decided to sell in a foreign country, it must determine the best mode of entry.
   b. Figure 15-2 shows three market entry strategies, along with the options each one offers.
   c. Each succeeding strategy involves more commitment and risk, but also more control and potential profits.

Exporting
   d. The simplest way to enter a foreign market is through exporting.
   e. The company may passively export its surpluses from time to time, or it may make an active commitment to expand exports to a particular market.
      1. In either case, the company produces all its goods in its home country.
      2. It may or may not modify them for the export market.
   f. Companies typically start with indirect exporting, working through independent international marketing intermediaries.
      1. Indirect exporting involves less investment because the firm does not require an overseas sales force or set of contacts.
      2. It also involves less risk.
   g. Sellers may eventually move into direct exporting.
      1. The investment and risk are somewhat greater in this strategy, but so is the potential return.
      2. A company can set up a domestic export department that carries out export activities.

3. The company can set up an overseas sales branch that handles sales, distribution, and perhaps promotion.

4. The company can also send home-based salespeople abroad at certain times in order to find business.

5. The company can do its exporting either through foreign-based distributors who buy and own the goods or through foreign-based agents who sell the goods on behalf of the company.

Joint Venturing

h. Joint venturing occurs when the company joins with foreign companies to produce or market products or services.

    1. It differs from exporting in that the company joins with a host country partner to sell or market abroad.

    2. It differs from direct investment in that an association is formed with someone in the foreign country.

i. There are four types of joint ventures.

    1. Licensing is a simple way for a manufacturer to enter international marketing.

        i. For a fee or royalty, the licensee buys the right to use the company's manufacturing process, trademark, patent, trade secret, or other item of value.

        ii. The company thus gains entry into the market at little risk; the licensee gains production expertise or a well-known product or name without having to start from scratch.

        iii. Disadvantages include the firm having less control over the licensee than it would over its own production facilities; if the licensee is very successful, the firm has given up these profits, and if and when the contract ends, it may find it has created a competitor.

    2. In contract manufacturing, the company contracts with manufacturers in the foreign market to produce its product or provide its service.

        i. The drawbacks of contract manufacturing are decreased control over the manufacturing process and loss of potential profits on manufacturing.

        ii. The benefits are the chance to start faster, with less risk, and the later opportunity either to form a partnership with or to buy out the local manufacturer.

    3. Under management contracting, the domestic firm supplies management know-how to a foreign company that supplies the capital.

        i. Management contracting is a low-risk method of getting into a foreign market, and it yields income from the beginning.

        ii. The arrangement is not sensible if the company can put its scarce management talent to better uses or if it can make greater profits by undertaking the whole venture.

   iii. Management contracting also prevents the company from setting up its own operations for a period of time.

  4. Joint ownership ventures consist of one company joining forces with foreign investors to create a local business in which they share joint ownership and control.

   i. A company may buy an interest in a local firm, or the two parties may form a new business venture.

   ii. Joint ownership may be needed for economic or political reasons.

   iii. The firm may lack the financial, physical, or managerial resources to undertake the venture alone.

   iv. Or a foreign government may require joint ownership as a condition for entry.

   v. The drawbacks include the fact that the partners may disagree over investment, marketing, or other policies.

Direct Investment

j. The biggest involvement in a foreign market comes through direct investment—the development of foreign-based assembly or manufacturing facilities.

k. Foreign product facilities offer many advantages.

  1. The firm may have lower costs in the form of cheaper labor or raw materials, foreign government investment incentives, and freight savings.

  2. The firm may improve its image in the host country because it creates jobs.

  3. Generally, a firm develops a deeper relationship with government, customers, local suppliers, and distributors.

  4. The firm keeps control over the investment and therefore can develop manufacturing and marketing policies that serve its long-term international objectives.

l. The main disadvantage of direct investment is that the firm faces many risks, such as restricted or devalued currencies, falling markets, or government changes.

7. **Deciding on the Global Marketing Program**

a. Companies that operate in one or more foreign markets must decide how much, if at all, to adapt their marketing mixes to local conditions.

  1. At one extreme is a standardized marketing mix, selling largely the same products and using the same marketing approaches world-wide.

  2. At the other extreme is an adapted marketing mix. In this case, the producer adjusts the marketing mix elements to each target market, bearing more costs but hoping for a larger market share and return.

b. Global standardization is not an all-or-nothing proposition but rather a matter of degree.
  1. Companies should look for ways to standardize to help keep down costs and prices and to build greater global brand power.
  2. But they must not replace long-run marketing thinking with short-run financial thinking. Although standardization saves money, marketers must make certain that they offer what consumers in each country want.
  3. Many possibilities exist between the extremes of standardization and complete adaptation.
  4. Most international marketers suggest that companies should "think globally but act locally"—that they should seek a balance between standardization and adaptation. These marketers advocate a "glocal" strategy in which the firm standardizes certain core marketing elements and localizes others.

Product

c. Five strategies allow for adapting product and promotion to a global market. See Figure 15-3. Three are product strategies and two are communication strategies.

  1. Straight product extension means marketing a product in a foreign market without any change.
     i. The first step should be to find out whether foreign consumers use that product and what form they prefer.
  2. Product adaptation involves changing the product to meet local conditions or wants.
     i. In some instances, products must be adapted to local customs or spiritual beliefs.
  3. Product invention consists of creating something new for a specific country market. This strategy can take two forms.
     i. It might mean reintroducing earlier product forms that happen to be well adapted to the needs of a given country.
     ii. Or a company might create a new product to meet a need in a given country.

Promotion

d. Companies can either adopt the same promotion strategy they used in the home market or change it for each local market.
e. Some global companies use a standardized advertising theme around the world.
f. Colors may need to be changed to avoid taboos in other countries.
g. Some companies use communication adaptation, which is fully adapting their advertising messages to local markets.
h. Media also need to be adapted internationally because media availability varies from country to country.

<u>Price</u>

i.   Companies also face many problems in setting their international prices.

j.   Foreign prices will be higher than domestic prices.

        1.   Price escalation occurs because of added costs of transportation, tariffs, importer margin, wholesaler margin, and retailer margin. Depending on these costs, the product may have to sell for two to five times as much in another country to make the same profit.

k.   Setting prices for goods that a company ships to its foreign subsidiaries can be problematic.

        1.   If a company charges a foreign subsidiary too much, it may end up paying higher tariff duties even while paying lower income taxes in that country.

        2.   If the company charges its subsidiary too little, it can be charged with dumping.

                i.   Dumping occurs when a company either charges less than it costs or less than it charges in its home market.

        3.   Recent economic and technological forces have had an impact on global pricing.

                i.   In the European Union the transition to the euro is reducing the amount of price differentiation.

                ii.   When firms sell their wares over the Internet, customers can see how much products sell for in different countries.

                iii.   People can order directly from the company location or dealer offering the lowest price.

                iv.   This will force companies toward more standardized international pricing.

<u>Distribution Channels</u>

l.   The international company must take a whole-channel view of the problem of distributing products to final consumers. Figure 15-4 shows the three major links between the seller and the final buyer.

        1.   The seller's headquarters organization supervises the channels and is part of the channel itself.

        2.   The channels between nations move the products to the borders of the foreign nations.

        3.   The channels within nations move the products from their foreign entry point to the final consumers.

m.   Channels of distribution vary greatly from nation to nation.

        1.   There are large differences in the numbers and types of inter-mediaries serving each foreign market.

        2.   The size and character of the retail units abroad also differ.

8.   Deciding on the Global Marketing Organization

a.   Companies manage their international marketing activities in at least three different ways.

        1.   A firm normally gets into international marketing by simply shipping out its goods.

    i. If its international sales expand, the company organizes an export department with a sales manager and a few assistants.

    ii. As sales increase, the export department can expand to include various marketing services so that it can actively go after business.

  2. An international division or subsidiary will be formed when companies get involved in several international markets and ventures. They can be organized in several ways.

    i. A geographic organization has country managers who are responsible for salespeople, sales branches, distributors, and licensees in their respective countries.

    ii. World product groups are each responsible for worldwide sales of different product groups.

    iii. International subsidiaries are each responsible for their own sales and profits.

  3. In a truly global organization, the company stops thinking of itself as a national marketer who sells abroad and starts thinking of itself as a global marketer.

    i. The top corporate management and staff plan worldwide manufacturing facilities, marketing policies, financial flows, and logistical systems.

    ii. The global operating units report directly to the chief executive or executive committee of the organization, not to the head of an international division.

Creative Marketing Exercises Designed to Reinforce the Concepts!!! (Suggested answers to these exercises can be found at the end of the Study Guide.)

1. Locate a governmental study on how foreign imports affect U.S. jobs. Pinpoint specific statistics for your state.

2. Go to www.wto.org and locate a list of countries that have signed GATT so far. Who is missing?

3. What is the primary difference between the European Union and GATT? Find and print a list of the fundamental rights of the members of the EU, using your favorite search engine.

4. List the primary factors involved in determining duty rates. Use www.customs.gov as a resource.

5. Make a list of the natural resources of Iran. What is its primary export? Import? How could this information be a useful marketing tool?

6. Locate an online currency converter and calculate the value of $1,000 in France, Germany, Spain, Bolivia, China, and South Africa.

7. Go to www.export.gov and determine the number 1 export for the United States.

8.   You have been asked to create a baby doll for sale to children in the Middle East. What features would be different from that of an American doll?

9.   Identify 5 examples of current products that would have to change something about the packaging in order to be sold in a foreign market. Justify your selection by listing the product, the country in which it is meant to be sold, and the cultural issue that needs to be addressed.

10. Go to www.ebay.com and search for products sold by international sellers. What types of products are prevalent?

"Linking the Concepts" – Suggestions/Hints

1. McDonald's should standardize their signature products such as hamburgers, fries, and soft drinks. This would need to be done so consumers would recognize this global company. The soft drinks would be easy due to the fact that Coca-Cola is already well known and it is the soft drink of McDonald's. McDonald's will need to develop some localized products, as well. These products will need to reflect the local tastes of their consumers.

    One successful marketing strategy/program that would be successful in China would be a tie-in to their national basketball team. There has been an increase in the interest, internationally, of their team and players. Some have even gone on to play in the NBA. Here is a chance to tie in an American product to an American game in hopes of creating interest in McDonald's.

2. In terms of the Canadian market, it is easier to standardize McDonald's products. The tastes of the Canadian market are very similar to the American market. McDonald's would find it easy to introduce more items off the American menu into the Canadian market. Their signature products, (Big Mac, fries, shakes) should be standardized and sold in the Canadian market. Again, this would allow for global recognition. McDonald's should also develop products that are of interest to the local markets, as well. This would help show the local market that they understand them as consumers.

    Marketing programs in Canada would be successful if tied in to sports, as well. Hockey is the major sport in Canada and marketing programs associated with this sport would prove successful.

3. A lot of McDonald's globalization programs have contributed to the "Americanization" of international sites. McDonald's restaurants have created a sense of visiting America without leaving their country, for some consumers. A lot of Americans look for a McDonald's so they can get a taste of home. However, because McDonald's is so American, groups of terrorists have attacked the restaurants in an effort to show the world that they have, in a way, attacked America.

Marketing Adventure Exercises (Suggested answers to these exercises can be found at the end of the Study Guide.)

(Visit www.prenhall.com/adventure for advertisements.)

1.   Auto   Student choice
      Food   Student choice
      Nonprofit   Student choice

The text states that "most international advertisers 'think globally but act locally'." Discuss this strategy in relation to ads in the selected categories.

2.   Auto   Ford

Consider Ford as a company in the Auto category that sells products globally. Discuss the economic and legal risks that Ford can face in the global marketplace.

3.   Apparel   Levis

What are the questions that Levi's had to answer as it moved toward globalization?

4.   Student choice

When this product is sold in foreign countries, there are various trade restrictions facing a company. Define these restrictions and discuss which category might be likely to face such restrictions.

5.   Cosmetics   Clearblue, Hawaiian Tropic

In an earlier chapter you learned about the macroenvironment that a firm faces when conducting marketing activities. This includes the large societal forces that affect an organization including demographic, economic, natural, technological, political, and cultural forces. Review the selected ads and then discuss the macroenvironmental factors that you believe affect these products in the global market.

6.   Food-Beverage   Student choice

As discussed in your text, some critics of globalization have said that "globalization really means "Americanization." Are there ads for products in the Food category for which this may be true?

7.   Household   Ikea

Ikea had to decide whether it wanted to compete in the international marketplace. What were the advantages and disadvantages it faced?

8.   Travel   Quantas, Virgin Atlantic

Marketers must decide whether to adapt products and advertisements to meet the unique cultures and needs of consumers across international borders or to standardize the marketing mix across those borders. Match the selected ads to the type of strategy they employ and discuss which is more effective.

9.   Student choice

Discuss the differences between a straight product extension, a product adaptation, and a product invention. Find ads with products that serve as examples.

10.   Food   Pullman, Snickers

Companies can adopt the same promotion strategy they used in the home market or change it for each local market. Discuss the strategies used for Snickers candy bar and Pullman bread.

SUM IT UP!!!!!!

Using only this page, sum up all of the concepts and terms discussed in Chapter 15 – "The Global Marketplace". Here is your chance to make sure you know and understand the concepts!!!!

Chapter 16
Marketing Ethics and Social Responsibility

Previewing the Concepts: Chapter Objectives

1. Identify the major social criticisms of marketing.
2. Define *consumerism* and *environmentalism*, and explain how they affect marketing strategies.
3. Describe the principles of socially responsible marketing.
4. Explain the role of ethics in marketing.

JUST THE BASICS

Chapter Overview

There have been many criticisms of marketing as it impacts individual consumers, other businesses, and society as a whole. This chapter discusses them in detail and provides some responses to critics.

Surveys usually show that consumers hold mixed or even slightly unfavorable attitudes toward marketing practices. Consumers, consumer advocates, government agencies, and other critics have accused marketing of harming consumers through high prices, deceptive practices, high-pressure selling, shoddy or unsafe products, planned obsolescence, and poor service to disadvantaged consumers. Companies respond in many ways, noting that the practices that cause so-called high prices—advertising, using intermediaries, decorative packaging, and the like—are necessary and well liked by consumers. Companies also note that those who use deceptive practices and high-pressure selling, or produce shoddy products, will not remain in business long.

Critics also charge that the American marketing system has added to several "evils" in society at large. Advertising has been a special target; it is charged that it urges too much interest in material possessions. However, our wants and values are influenced not only by marketers, but also by family, peer groups, religion, ethnic background, and education. If Americans are materialistic, these values arose out of basic socialization processes that go much deeper than business and mass media could produce alone.

Consumerism is an organized movement of citizens and government agencies to improve the rights and powers of buyers in relation to sellers. Consumers not only have the right but also the responsibility to protect themselves instead of leaving this function to someone else.

Environmentalism is an organized movement of concerned citizens, businesses, and government agencies to protect and improve people's living environments.

Environmentalists are not against marketing and consumption; they simply want people and organizations to operate with more care for the environment. Companies have responded. At the most basic level, a company can practice pollution prevention, which means eliminating or minimizing waste before it is created. At the next level, companies can practice product stewardship, which is minimizing not just pollution from production but all environmental impacts through the full product life cycle. At the third level of environmental sustainability, companies look to the future and plan for new environmental technologies. Finally, companies can develop a sustainability vision, which serves as a guide to the future.

The philosophy of enlightened marketing holds that a company's marketing should support the best long-run performance of the marketing system. It consists of five principles: consumer-oriented marketing, innovating marketing, value marketing, sense-of-mission marketing, and societal marketing.

Companies also need to develop corporate marketing ethics policies, which are broad guidelines that everyone in the organization must follow. These policies should cover distributor relations, advertising standards, customer service, pricing, product development, and general ethical standards. For the sake of all of the company's stakeholders—customers, suppliers, employees, shareholders, and the public—it is important to make a commitment to a common set of shared standards worldwide. Ethics and social responsibility require a total corporate commitment. They must be a component of the overall corporate culture.

Chapter Outline

1. **Introduction**
   a. Nike has been a lightning rod for social responsibility criticisms. Critics have accused Nike of putting profits ahead of the interests of consumers and the broader public, both at home and abroad.
   b. Despite its success at selling shoes, Nike has been accused of everything from running sweatshops, using child labor, and exploiting low-income consumers to degrading the environment.
   c. Nike outsources production to contractors in low-wage countries. It has created a Code of Conduct, which demands socially responsible labor practices by its contractors. Nike has actually been improving the working conditions in low-wage countries.
   d. Nike has also received criticism at home. It has been accused of inappropriately targeting its most expensive shoes to low-income families, making the shoes an expensive status symbol for poor urban kids.
   e. A closer look shows that Nike works hard at being a socially responsible global citizen. Nike and the Nike Foundation contributed more than $29 million in cash and products last year to programs that encourage youth to participate in sports and that address challenges of globalization.

f. Nike also donates money for education, community development, and small-business loans in the countries in which it operates.

g. Responsible marketers discover what consumers want and respond with marketing offers that give satisfaction and value to buyers and profit to the producer. The marketing concept is a philosophy of customer satisfaction and mutual gain.

h. But some companies use questionable marketing practices, and some marketing actions that seem innocent in themselves strongly affect the larger society.

2. **Social Criticisms of Marketing**

a. Social critics claim that certain marketing practices hurt individual consumers, society as a whole, and other business firms.

Marketing's Impact on Individual Consumers

b. Surveys usually show that consumers hold mixed or even slightly unfavorable attitudes toward marketing practices.

c. Many critics charge that the American marketing system causes prices to be higher than they would be under more "sensible" systems.

1. A long-standing charge is that greedy intermediaries mark up prices beyond the value of their services. There are too many intermediaries, they are inefficient, or they provide unnecessary or duplicate services.

    i. Companies respond that intermediaries do work that would otherwise have to be done by manufacturers or consumers. Markups reflect services that consumers themselves want—more convenience, larger stores and assortments, longer store hours, return privileges, and others.

2. Modern marketing is also accused of pushing up prices to finance heavy advertising and sales promotion. Much of the packaging and promotion adds only psychological value to the product rather than functional value.

    i. Marketers respond that consumers can usually buy functional versions of product at lower prices. However, they are willing to pay more for products that also provide psychological benefits.

    ii. Heavy advertising and promotion may be necessary for a firm to match competitors' efforts---the business would lose "share of mind" if it did not match competitive spending.

3. Critics also charge that some companies mark up goods excessively.

    i. Marketers respond that most businesses try to deal fairly with consumers because they want repeat business.

    ii. Marketers also respond that consumers often don't understand the reasons for high markups.

d. Marketers are sometimes accused of deceptive practices that lead consumers to believe they will get more value than they actually do. However, marketers argue that most companies avoid deceptive practices because such practices harm their businesses in the long run. Deceptive practices fall into three groups.

    1. Deceptive pricing includes practices such as falsely advertising "factory" or "wholesale" prices or a large reduction from a phony high retail list price.

    2. Deceptive promotion includes practices such as overstating the product's features or performance, luring the customer to the store for a bargain that is out of stock, or running rigged contests.

    3. Deceptive packaging includes exaggerating package contents through subtle design, not filling the package to the top, using misleading labeling, or describing size in misleading terms.

e. Salespeople are sometimes accused of high-pressure selling that persuades people to buy goods they had no thought of buying.

    1. Marketers know that buyers often can be talked into buying unwanted or unneeded things.

    2. In most cases, however, marketers have little to gain from high-pressure selling. Such tactics may work in one-time selling situations for short-term gain. However, most selling involves building long-term relationships with valued customers.

f. Another criticism is that products lack the quality they should have. Many products are not made well and services are not performed well. Many products deliver little benefit, or they might even be harmful.

    1. Product safety has been a problem for several reasons, including company indifference, increased product complexity, and poor quality control.

    2. Most manufacturers want to produce quality goods. The way a company deals with product quality and safety problems can damage or help its reputation.

g. Critics have also charged that some producers follow a program of planned obsolescence, causing their products to become obsolete before they actually should need replacement.

    1. Marketers respond that consumers like style changes; they get tired of old goods and want a new look in fashion or a new design in cars.

    2. Companies do not design products to break down early, because they do not want to lose customers to other brands. Instead, they seek improvement to ensure that products will consistently meet or exceed customer expectations.

h. The American marketing system has been accused of serving disadvantaged consumers poorly. Critics claim that the urban poor often have to shop in smaller stores that carry inferior goods and charge higher prices.

1. Better marketing systems must be built to service disadvantaged consumers. Disadvantaged consumers clearly need consumer protection.
2. The FTC has taken action against merchants who advertise false values, sell old merchandise as new, or charge too much for credit.

Marketing's Impact on Society as a Whole

i. The American marketing system has been accused of adding to several "evils" in American society at large. Advertising has been a special target.

j. Critics have charged that the market system urges too much interest in material possessions.
  1. The critics do not view this interest in material things as a natural state of mind but rather as a matter of false wants created by marketing.
  2. On a deeper level, our wants and values are influenced not only by marketers but also by family, peer groups, religion, ethnic background, and education.

k. Business has been accused of overselling private goods at the expense of public goods.
  1. As an example, an increase in automobile ownership (private good) requires more highways, traffic control, parking spaces, and police services (public goods).
  2. A way must be found to restore a balance between private and public goods.
     i. One option is to make producers bear the full social costs of their operations.
     ii. A second option is to make consumers pay the social costs.

l. Critics charge the marketing system with creating cultural pollution.
  1. Our senses are being constantly assaulted by advertising.
  2. Marketers answer that they hope their ads reach the primary target audience. Also, ads make much of television and radio free to users and keep down the costs of magazines and newspapers. Finally, today's consumers have alternatives.

m. Another criticism is that business wields too much political power.
  1. Advertisers are accused of holding too much power over the mass media, limiting their freedom to report independently and objectively.
  2. American industries do promote and protect their own interests. They have a right to representation in Congress and the mass media, although their influence can become too great.

Marketing's Impact on Other Businesses

n. Critics also charge that a company's marketing practices can harm other companies and reduce competition.

1. Critics claim that firms are harmed and competition is reduced when companies expand by acquiring competitors rather than by developing their own new products.
2. Critics have also charged that marketing practices bar new companies from entering an industry.
3. Finally, some firms have in fact used unfair competitive marketing practices with the intention of hurting or destroying other firms.
4. Various laws work to prevent such predatory competition. It is difficult, however, to prove that the intent or action was really predatory.

3. **Citizen and Public Actions to Regulate Marketing**
   a. Grassroots movements have arisen from time to time to keep businesses in line.

Consumerism
   b. American business firms have been the target of organized consumer movements on three occasions.
      1. The first consumer movement took place in the early 1900s. It was fueled by rising prices, Upton Sinclair's writings on conditions in the meat industry, and scandals in the drug industry.
      2. The second consumer movement, in the mid-1930s, was sparked by an upturn in consumer prices during the Great Depression and another drug scandal.
      3. The third movement began in the 1960s. Consumers had become better educated, products had become more complex and potentially more hazardous, and people were unhappy with American institutions.
   c. Consumerism is an organized movement of citizens and government agencies to improve the rights and powers of buyers in relation to sellers.
      1. Traditional sellers' rights include:
         i. The right to introduce any product in any size and style, provided it is not hazardous to personal health or safety; or, if it is, to include proper warnings and controls.
         ii. The right to charge any price for the product, provided no discrimination exists among similar kinds of buyers.
         iii. The right to spend any amount to promote the product, provided it is not defined as unfair competition.
         iv. The right to use any product message, provided it is not misleading or dishonest in content or execution.
         v. The right to use any buying incentive programs, provided they are not unfair or misleading.
      2. Traditional buyers' rights include:
         i. The right not to buy a product that is offered for sale.
         ii. The right to expect the product to be safe.
         iii. The right to expect the product to perform as claimed.

3. Consumer advocates call for the following additional consumer rights:
    i. The right to be well informed about important aspects of the product.
    ii. The right to be protected against questionable products and marketing practices.
    iii. The right to influence products and marketing practices in ways that will improve the "quality of life."
d. Consumers have not only the right but also the responsibility to protect themselves instead of leaving this function to someone else.

Environmentalism
e. Environmentalism is an organized movement of concerned citizens, businesses, and government agencies to protect and improve people's living environments.
f. Environmentalists are not against marketing and consumption; they simply want people and organizations to operate with more care for the environment.
g. The first wave of modern environmentalism in the United States was driven by environmental groups and concerned consumers in the 1960s and 1970s.
    1. They were concerned about the damage to the ecosystem caused by strip-mining, forest depletion, acid rain, loss of the atmosphere's ozone layer, toxic wastes, and litter.
h. The second environmentalism wave was driven by government, which passed laws and regulations during the 1970s and 1980s governing industrial practices that have an impact on the environment.
i. Those two movements are now merging into a third and strong wave in which companies are accepting responsibility for doing no harm to the environment.
    1. More and more companies are adopting policies of environmental sustainability—developing strategies that both sustain the environment and produce profits for the company.
j. Figure 16-1 shows a grid that companies can use to gauge their progress toward environmental sustainability.
    1. At the most basic level, a company can practice pollution prevention.
    2. At the next level, companies can practice product stewardship—minimizing not just pollution from production but all environmental impacts through the full product life cycle.
        i. Many companies are adopting design for environment (DFE) practices, which involve thinking ahead in the design stage to create products that are easier to recover, reuse, or recycle.
    3. At the third level of environmental sustainability, companies look to the future and plan for new environmental technologies.

i.   Many organizations that have made good headway in pollution prevention and product stewardship are still limited by existing technologies.

4.   Finally, companies can develop a sustainability vision, which serves as a guide to the future. It shows how the company's products and services, processes, and policies must evolve and what new technologies must be developed to get there.

k.   Environmentalism creates some special challenges for global marketers.

1.   As international trade barriers come down and global markets expand, environmental issues are having an ever-greater impact on international trade.

2.   Environmental policies still vary widely from country to country, and uniform worldwide standards are not expected for many years.

Public Actions to Regulate Marketing

l.   Figure 16-2 illustrates the major legal issues facing marketing management.

4.   Business Actions Toward Socially Responsible Marketing

a.   Most companies have grown to embrace the new consumer rights, at least in principle. They might oppose certain pieces of legislation as inappropriate ways to solve specific consumer problems, but they recognize the consumer's right to information and protection.

Enlightened Marketing

b.   The philosophy of enlightened marketing holds that a company's marketing should support the best long-run performance of the marketing system. It consists of five principles.

1.   Consumer-oriented marketing means that the company should view and organize its marketing activities from the consumer's point of view. It should work hard to sense, serve, and satisfy the needs of a defined group of customers.

2.   The principle of innovative marketing requires that the company continuously seeks real product and marketing improvements. The company that overlooks new and better ways to do things will eventually lose customers to another company that has found a better way.

3.   According to the principle of value marketing, the company should put most of its resources into value-building marketing investments.

4.   Sense-of-mission marketing means that the company should define its mission in broad social terms rather than narrow product terms. When a company defines a social mission, employees feel better about their work and have a clearer sense of direction.

5.   Following the principle of societal marketing, an enlightened company makes marketing decisions by considering consumers'

wants and interests, the company's requirements, and society's long-run interests.

c. Figure 16-3 shows how products can be classified according to their degree of immediate consumer satisfaction and long-run consumer benefit.

1. Deficient products, such as bad-tasting and ineffective medicine, have neither immediate appeal nor long-run benefits.

2. Pleasing products give high immediate satisfaction but may hurt consumers in the long run.

3. Salutary products have low appeal but may benefit consumers in the long run.

4. Desirable products give both high immediate satisfaction and high long-run benefits.

d. Companies should try to turn all of their products into desirable products.

Marketing Ethics

e. Companies need to develop corporate marketing ethics policies—broad guidelines that everyone in the organization must follow.

1. These policies should cover distributor relations, advertising standards, customer service, pricing, product development, and general ethical standards.

f. Table 16-1 lists some difficult ethical situations marketers could face during their careers.

g. But what principles should guide companies and marketing managers on issues of ethics and social responsibility?

1. One philosophy is that such issues are decided by the free market and legal system. Under this principle, companies and their managers are not responsible for making moral judgments. Companies can in good conscience do whatever the market and legal systems allow.

2. A second philosophy puts responsibility not on the system but in the hands of individual companies and managers. This more enlightened philosophy suggests that a company should have a "social conscience."

h. Under the societal marketing concept, each manager must look beyond what's legal and allowed and develop standards on personal integrity, corporate conscience, and long-run consumer welfare.

i. As with environmentalism, the issue of ethics provides a special challenge for international marketers.

1. Business standards and practices vary a great deal from one country to the next.

2. For the sake of all the company's stakeholders—customers, suppliers, employees, shareholders, and the public—it is important to make a commitment to a common set of shared standards worldwide.

j. Many industrial and professional associations have suggested codes of ethics, and many companies are now adopting their own codes.

       1.   Table 16-2 shows the code of ethics for the American Marketing Association.

   k.   Ethics and social responsibility require a total corporate commitment. They must be a component of the overall corporate culture.

Creative Marketing Exercises Designed to Reinforce the Concepts!!! (Suggested answers to these exercises can be found at the end of the Study Guide.)

1.   Make an argument for or against an additional sales tax on alcohol. Do you see this as a punishment or a way to generate revenue? Explain.

2.   Does your college bookstore have a captive audience in you? Is it an ethical issue for you that the bookstore can determine the price of books without justification? Explain.

3.   Go to www.congress.gov and conduct a background search on the Wheeler-Lea Act. Outline the circumstances that preceded this bill.

4.   Debate the ethical issue of a "pyramid scheme." Should these be illegal? Why or why not?

5.   How did Ralph Nader become a fierce consumer advocate? Give a brief biographical background.

6.   Formulate a list of 5 organized consumer movements and outline each group's primary mission.

7.   List 5 websites that you consider to be examples of "green marketing."

8.   Go to www.enron.com and find a copy of the company's "Code of Ethics." Give 5 examples of how the fall of Enron can be contributed to specific breaks in its code.

9.   Go to the website for your state legislature. How many bills are currently being considered under the title of "ethics"?

10.   You are the CEO of a baby food company and have discovered that 100 cases of baby food may contain a mild form of salmonella. You shipped 1,000,000 cases of product last week alone. What do you do?

"Linking the Concepts" (#1) – Suggestions/Hints

1.   A couple of examples of marketing that have been abused are telemarketing and some websites on the Internet. These two examples have had numerous new stories presented on how they have "ripped off" consumers. There have been many cases of telemarketers who have bilked the elderly out of their life savings. Some Internet sites have promised goods and services, but they never arrived. There are also numerous credit card fraud cases based on the Internet. These examples have all been intentional in the sense that they were designed to fraud customers. The common thread through each of these examples is that the "salespeople" were marketing to consumers using a type of media that did not

require face-to-face meetings with their victims. They were able to hide behind the phone or the Internet. Because the victims did not know their true sources of the contact, it made it hard for them to prosecute. The criminals were able to set up shop, steal, and then quickly disband.

2. One way to avoid Internet fraud would be to just use the Internet as a source of information, not as the final retail destination. Look up on the Internet what you are shopping for, then call them for brochures or visit their place of business. See if the store is a real location. Try to require face-to-face transactions. If you cannot actually go to the store, contact the Better Business Bureau to see if they are a legitimate retail establishment. Also, if you cannot visit the store, ask for a catalog or paperwork to be mailed to your home. This way you can examine it. Be careful about "blindly" ordering stuff from the Internet. Do you know the source? Pay Pal accounts on the Internet also help in preventing fraud. Some of the big auction sites on the Internet use Pay Pal accounts to help their customers feel safe.

"Linking the Concepts" (#2) – Suggestions/Hints

1. The philosophy of enlightened marketing holds that a company's marketing should support the best long-run performance of the marketing system. All of the elements of a company's marketing mix should be customer-oriented, have innovativeness, value, sense of mission, and the concerns of society.

2. McDonald's is a good example of a company when examining the five principles of enlightened marketing. The following is a quick look at the five principles:

Customer-oriented: McDonald's is constantly trying to develop products and services that meet customers' changing needs and tastes. They are always trying to develop new sandwiches to meet those needs.

Innovative marketing: McDonald's is very aggressive at trying to come up with new ways to market and promote their products. They are trying new games, new commercials, and new movie tie-ins all the time.

Value marketing: McDonald's has been very successful at developing value-added menu meals. They try to keep their prices low and offer quality products and services to their customers.

Sense-of-mission: McDonald's has worked hard to try to become the global company that they are, without sacrificing their mission of delivering quality products and services. They have stayed very true to their original mission of giving back to the communities. This can be proven by the success of the Ronald McDonald Children's Charities Programs.

Societal marketing: McDonald's has become the benchmark for a lot of companies as to how to give back to the community.   They are involved in nutritional programs, children's programs, educational programs, and youth sports programs, as examples.

3.   Philip Morris could be considered an example of a company that has not followed the enlightened marketing principles. They are good at offering customer-oriented marketing because they are giving their target markets what they want. They are innovative in the way they present their products. They have stayed current as to technology while being innovative. They offer value to their target markets and they have followed the mission of the company.
However, there has been a major debate over the years as to how they are concerned about the values and health of society in general.

Marketing Adventure Exercises (Suggested answers to these exercises can be found at the end of the Study Guide.)

(Visit www.prenhall.com/adventure for advertisements.)

1.   Student choice

Critics of marketing cite high advertising and promotion costs for higher consumer prices. Select ads that promote types of products that are heavily advertised and may be marked up 40% or so.

2.   Cosmetics   Celebrex

Excessive markups can be found in the pharmaceutical industry, according to some critics. If you were the marketing manager for the selected product, what would your response be?

3.   Apparel   Student choice
        Electronics   Student choice

Discuss planned obsolescence, and then select ads that feature products that are quickly outdated.

4.   Auto   Student choice

The text mentions that social scientists believe that today society has abandoned the materialistic wants of the 1980's, and has a greater commitment to society and an emphasis on more basic values. Review the ads in the Auto category to find examples supporting this claim and others that refute it.

5.   Food   Barilla, Rianxeira

Discuss consumerism and how it relates to the selected product.

6.   Student choice

Today's marketers must take greater responsibility for the social and environmental impacts of their actions.   Find an ad that shows such responsibility and then visit the company's web site to read its mission statement.

7.   Auto   Toyota
      Cosmetics   Hawaiian Tropic
      Financial   Storebrand

According to the text, products can be classified according to their degree of immediate consumer satisfaction and long-run consumer benefit. List these classifications and decide where the products in the selected ads belong.

8.   Student choice

Earlier in the text it was suggested that "enlightened companies encourage their managers" to go beyond what the law allows and just "do the right thing." Choose an ad that shows an enlightened firm that has shown socially responsible behavior in terms of its products.

SUM IT UP!!!!!!

Using only this page, sum up all of the concepts and terms discussed in Chapter 16 – "Marketing Ethics and Social Responsibility". Here is your chance to make sure you know and understand the concepts!!!!

Answers to the "Creative Marketing Exercises Designed to Reinforce the Concepts!!!"

Chapter 1

1.   List 5 items that you perceive as "needs." What differentiates them from "wants?" Explain.

      The purpose here is to distinguish between needs and wants. One solution will be to list things that are thought of as needs such as a computer, new car, new stereo, video games, etc. The discussion should focus on these items as wants, not needs.

2.   Think of a product you were "offered" recently. Outline the manner in which it was presented to you and discuss why you did or did not accept this offer.

      You may observe a billboard, TV or print ad, or a display in a store window. The observation could be a bag of chips in a vending machine. The point is that products are offered in numerous ways in order to meet the needs and wants of millions of customers.

3.   A friend offers to make a trade with you. He offers you the guitar of your dreams for something of yours. What would you offer in exchange and why?

      You may substitute the guitar for something you truly value. The point of this question is to get you to understand that you will exchange something for something else that you value. The most common exchange tool is money, but money is not the only thing perceived as valuable.

4.   Select your favorite frozen food. To whom do you think this product is targeted? Explain.

      You may select a product like "Lean Cuisine" and discuss that this product is designed for those diet conscious people that need to eat quickly. Working women are the primary target. "Hearty man" frozen dinners on the other hand would target single men that need something hearty and quick to eat.

5.   You have created a new software program that will streamline the payroll process for small businesses. Which of the 5 marketing concepts will you use to design and sell your product? Justify.

      You should pick one of the 5 concepts from page 12 and rationalize why your choice is the best pick for the software/small business market.

6.   Think about your favorite video store and the last experience you had while there. What makes this store your favorite?

      One solution might be to mention that you know the clerks in the store or are familiar with the layout of the videos. You may like the location, the parking situation, or the pricing methods. You will probably mention a combination of these.

7.   Find 5 examples of companies partnering to sell products.

      Convenience stores are great examples with Arby's, Subway, Taco Bell, Krispy Kreme, etc. represented in these venues. WalMart has several partnerships also.

8. How does your bank customize their services to meet the needs of its customers? Show examples.

   You could bring in brochures showing all the different types of accounts your bank offers.

9. Buick has a reputation for building customer loyalty from a specific demographic. Who are these customers and why are they so invested in Buick?

   Buick is known for selling cars to the senior citizen. Buick has been around for years and has built a reputation for quality and reliability.

10. Go to www.cocacola.com and read the company's international mission statement. Who are its target customers?

    This statement brings home the fact that Coke is a product for every human on the planet!

Chapter 2

1. Create a strategic plan for the next 5 years of your life. Determine where you are now, where you want to be in the future, and how you plan on getting there.

   Think about what you are doing now in your life and what you would like to be doing. Make a list of the hurdles you must jump to get to the finish line. Think of things like family, money, logistics, education, etc.

2. Find a copy of a mission statement for a dot.com company. Do you think its statement truthfully reflects its business practices and attitude toward customers? Explain.

   Look at Amazon, eBay, Yahoo, AOL Time Warner, etc. and compare the service they say they offer with the service they actually provide. Ask your friends for their opinions about online services.

3. Visit www.unicef.com and read the organization's mission statement. What is the vision of top management for this organization?

   Unicef offers a visionary look at the works of its organization. It wants to help those in need in all areas of the world and its mission and vision outlines everything from financial needs to emotional support.

4. Identify a company that hosts several strategic business units (SBUs). What is the logic for mixing those particular organizations together? Explain.

   Many companies now have several strategic business units to help generate revenue. Look at companies like Wal-Mart, Microsoft, Time Warner, and others.

5. Make a list of 3 companies that meet the description of each of the types of units in the Boston Consulting Group matrix. You should have 12 companies total.

   You can make an argument for almost any company to fit into any of these categories. Focus on the ability of the company to grow and generate revenue.

6.	Go to www.marriott.com and discuss how this organization has successfully diversified the hotel room.

Marriott offers several different brands to meet the needs of its customers. Courtyard, Residence Inn, and Fairfield are examples of a business class, extended-stay and budget offering from Marriott.

7.	Visit www.fritolay.com and list all the products in the snack value chain.

Lay's, Fritos, Doritos, Tostitos, Ruffles, Rold Gold, Cracker Jacks, Sun Chips, Santitas, WOW!, Grandma's, Munchos, and Funyuns

8.	Describe a local company that uses segmentation to sell its products. Is it successful?

Look at a rental car agency or a restaurant that offers an early bird special. These companies are pulling in different customers for the same product.

9.	You have been hired to develop the marketing plan for a new antique store in your community. How would you begin this process?

Use the marketing analysis on page 58. Start with planning, implementation, and end with control.

10.	The company you work for wants you to figure out why customer sales have dropped in the past few months. Use the marketing control process to speculate why there has been a decline.

Use Figure 2.7.

Chapter 3

1.	Identify the factors in your classroom environment that affect your learning performance.

Be sure to consider the desks, the lighting, the podium, and the arrangement of the seats. Other things to look at are the demeanor of the teacher and the students, the color of the walls, and the artwork or bulletin boards that may be distracting.

2.	Make a list of products that target senior citizens. Justify your choices.

A few choices would be muscle rub, vitamins, low sodium products, certain medications (arthritis, Alzheimer's drugs), frozen dinners, and various health aids, such as wheel chairs and electric beds and chairs.

3.	Think about your favorite reality television show. Why do you think these shows have become so popular?

Shows like "Joe Millionaire," "Survivor," and MTV's "The Osbournes" all target a different audience. Networks are claiming that the popularity of these shows stems from a desire to see how other people really live so that we can experience something other than our own lives.

4. Conduct a demographic survey of your class. Use gender, age, race, ethnic background, and marital status as your starting point.

Be sure to notice all the differences in your classmates. Students look at themselves as the same but when put in groups, the differences are startling.

5. Make a list of the members in your family that would be considered the "media public." Justify your choices.

Think about all the family members that buy something because they saw it on television or use a product because they read about it in a magazine.

6. List five characteristics of a Generation Xer. How does the entertainment industry appeal to this market? Give examples.

Consider all the young stars of music, movies, and television that try to appeal to a specific audience. Also, look at television shows that are branching out to many difference audiences.

7. Create a spreadsheet depicting your budget for this week. How does the amount you have left over affect your weekend plans?

Simply put, the more money you have left over, the more flexible you are to spend. The less you have, the tighter you tend to be with it.

8. Go to www.restaurant.org and locate the data depicting trends in eating out. What inference can you draw from this information?

Trends show people eating out more and more over the next 3 years. There is a definite trend in meal replacement and eating out for celebrations.

9. Visit the self-checkout aisle at your local grocery store. How does this technology better serve the customer?

Some would argue that the self-check system defeats the purpose of going to the grocery store. The self-check supporters enjoy shorter lines and can pace themselves instead of waiting on someone else.

10. How might the war in Iraq affect the sale of American goods in that region?

Consider the popularity of Americans at this time in Iraq and think about how American-made products would represent something that the Iraqis would appreciate.

Chapter 4

1. How accurate is the U.S. Census? How could this process be more effective and accurate?

Consider the "old method" of conducting the census: door-to-door surveys. How accurate is the information people give to the census taker? What happens if no one is home? Is that home revisited? The newer method involves the telephone. What about those individuals that do not have a phone? Look at all the ways information can be collected and put together in a package that encompasses all walks of life.

2.   Make a list of all the restaurants in a 5-mile radius. How is this information useful?

        Break the restaurants into types and then into price ranges. This information might be useful in determining what types of advertising would be effective in these areas and the types of customers that might be likely to visit.

3.   Call a local hotel and ask them how they conduct customer surveys. Do they feel this information is useful to their business?

        Most hotel chains have a written survey they leave in the guest rooms for completion. Some hotels are now using the touch screen method. Be sure to ask about incentives to customers for completing the survey and about how the collected data is utilized.

4.   Find a copy of a mail questionnaire and outline its advantages and disadvantages.

        Customer surveys that either come in the mail or have a postage-paid sticker on the card are a common form of data collection. Some of the advantages would be easy accessibility for the customer and a larger number of customers the company can access. Some disadvantages include low return rates and inaccurate information.

5.   Visit www.landsend.com and locate the customer touch points on this website.

        Look for links to company contact information, customer service, a shopping cart, a wish list, and any other area where the company can personalize its service.

6.   Conduct an Internet search for sites with international marketing research information. Which sites did you find to be the most helpful?

        The search engine Google.com is an effective tool for locating useful websites. There are numerous sites to see, so as you are surfing, focus on the ease at which you are able to find useful information.

7.   Go to www.courttv.com and take the "13th Juror" survey. How "up to date" is this information?

        These questions change everyday but how often is the information updated? How do you know that there is not one person taking the survey over and over again, in an effort to sway "public opinion"?

8.   Is "dumpster-diving" an ethical practice? Why or why not?

        Some would argue that once it's in the garbage can, it's public property. If a company wanted to protect something, it would not throw it in the garbage can. Others would argue that even though it's garbage, it is someone else's ideas that you are trying to obtain and that is the same as infringement on intellectual property rights.

9.   List 5 misleading statements advertisers make about their products.

        Look for statements that make broad general promises like "Makes your smile brighter" or "Enhances your natural glow." Also look for subjective terms that could mean several things.

10.   Create a course satisfaction survey for your marketing class and administer. What was the outcome?

Be honest but gentle in creating the questions. Think about what information you need to collect and then create a survey that meets those needs specifically.   Remember, the person and course you are critiquing can give you a grade!

Chapter 5

1.   What impact do you think you have on the U.S. economy? Global economy? Explain.

Consider the amount of money you spend every week and multiply that times 52. Then, take that number and multiply it by 50 (number of years you will probably live past your current age.) Is the number bigger than you anticipated? You will spend a large amount of money in your lifetime!

2.   Look through your favorite magazine and make a list of those items that are considered "trendy."

Items you want to consider are those that have been around a while and go in and out of style. Skirt lengths, suit cuts, shoe styles, makeup colors, models of cars, and music trends should all be considered.

3.   Visit www.atkinsdiet.com and give a brief background on this controversial diet plan. At what point did this plan change from being a "fad" to a "trend?"

This plan has been around a long time. Be sure to notice that the popularity has grown over time. What do you think contributed to this growth pattern? Read the scientific information and associate it with past and current trends.

4.   Identify a radio station in your listening area that caters to a subculture. What specific strategies do they use?

Pick a station that plays a specific type of music and has a DJ that fits a specific demographic. Easy listening, heavy metal, and country music stations are good choices.

5.   Outline 5 ways your family influences your buying. Be specific.

Consider the influence those around you have on what you buy, when you buy it, and how much you spend. Did your mother instill the notion in you that name brands are always the best or generic brands are the same only cheaper?

6.   Conduct a brief interview with a new mother. Ask about her diaper preference and what persuaded her to choose this particular brand.

Most new mothers choose a brand of diaper that their parents used or a brand used by all their friends. Diapers tend to be a product women choose based on word of mouth.

7. Make a list of your favorite television shows and write a brief synopsis on the lifestyle they depict.

This list will vary from person to person, but the selection of network and cable programming is astonishing in the variety of lifestyles you can see on TV.

8. Go to www.checksunlimited.com and discuss how this site is an example of how advertisers appeal to a person's personality to sell a product.

Be sure to notice the various categories of check styles and the numerous patterns to choose from. Every personality is represented from playful and childlike to business professional.

9. Find 5 examples of advertisements that use perception to sell a product.

Look for ads that use people and settings that have absolutely nothing to do with the product. Feminine hygiene products, laxatives, and senior citizen products are good examples.

10. Using Figure 5.5, outline the process you used in purchasing the last piece of clothing you acquired.

Follow the steps outlined in the figure: need recognition, information search, evaluation of alternatives, purchase decision, and post purchase behavior.

Chapter 6

1. Identify the various segments used in the pasta industry. How can these segments become narrower? Explain.

Consider all the shapes and sizes of pasta when thinking about this question. The product is the same and the shape of the product is the only differentiation between the types of pasta. To narrow the niche, companies are putting out products such as spinach pasta and sun-dried tomato pasta. Look at the makers of Pasta-roni and Lipton for more specialty types.

2. Flip through your favorite magazine and locate products targeted to teens. How did you know they were the intended target market?

Ads targeting teenagers tend to be colorful and have recognizable faces and places showcased. Look at products such as athletic shoes, cosmetics, and video games for examples.

3. Evaluate your calendar and make a list of all the birthdays you celebrate in a year. What financial impact do these occasions have on your budget?

The average person buys one birthday present a month and spends an average of $10 - $15 per gift. The financial impact on your budget should vary based on your income but focus on ALL the occasions you have to spend money on during the year!

4. Go to www.sheraton.com and name the various segments targeted.

Look at the pictures of the customers on the site. Who all are they trying to target?

5.   How many brands of diet beverages can you name? Is there a gap in this market? If so, where?

There are over 20 diet drinks on the market. There have been discussions about doing away with regular carbonated beverages. Ask someone who does NOT drink diet beverages why they do not!

6.   Analyze the current toothbrush market. Who are the main competitors and why do you think this market is so competitive?

Every magazine is filled with ads for new toothbrushes. Electric, leveled bristles, and polishing pads are just a few of the ways the makers of toothbrushes are differentiating their products.

7.   Using your favorite search engine, identify an organization that impresses you with its social responsibility. What attributes attracted you to this organization? Explain.

Pick a cause that interests you and conduct a search on that topic. If you come across a company like Enron or Worldcom, you are headed in the wrong direction!

8.   You are the owner of a new jewelry store in your local shopping mall. What is your competitive advantage? Outline your plan.

Consider the competition and what they are doing to draw in customers. What can you offer that they cannot? Think of ways to make your service special.

9.   Identify 5 examples of companies that market their product to the individual. What makes this approach work? Explain.

The text offers several examples of individual marketers. Look at the blue jean market, toy manufacturers, and car makers.

10.   Which value proposition appeals to you the most and why? More for more, more for the same, same for less, less for much less, or more for less.

Each person's answer will vary but consider why you buy something. Do you buy it because of the value or because you want the item?

Chapter 7

1.   Call your local physician and ask what products they sell versus what products they offer.

Your challenge here is to differentiate between what people consider a service, or an intangible good, and a product, something you can inventory. Every market is selling something. The question is "what?"

2.   List the various "products" your college/university sells.

Consider everything from books, pencils, and notepads to the expertise of the faculty and the use of the school facilities.

248

3. Go to www.esteelauder.com and look at their skin care line. Identify five products that can be viewed as unsought.

   Unsought products are those things that people might not know they need but buy anyway. Things like eye cream, pore refiner, and facial masks should be considered.

4. Go to www.amazon.com and read a review of the book by Private Jessica Lynch. How might this book be an example of "person marketing?"

   Consider how this book changes or shapes the public opinion of Jessica Lynch. Is she simply telling a story or selling herself?

5. Visit www.georgia.org and determine how Georgia is positioning the state as a tourism destination.

   Georgia showcases a myriad of attractions and sites of interest for all ages. Consider the scope of their offerings. Would Georgia be a place to visit for everyone?

6. Go to www.bestbuy.com and compare the features on two brands of DVD players and determine which player is the better value.

   Best Buy carries a wide variety of DVD players. Look at all the products and select two to compare. What are you looking for in a player and what are you willing to pay for? Consider your needs versus your wants, and determine what makes something a "value."

7. Visit www.marykay.com and locate their "Time Wise" line of anti-aging products. How is this line an example of product expansion?

   Identify the core product and then list the products that stem from that one. You should find that all the products have similar characteristics but are subtly different to accommodate the needs of different customers.

8. Visit www.polo.com and outline how Ralph Lauren has kept his product lines consistent with his mission.

   Ralph Lauren does not offer a huge variety of styles and colors. His products are traditional and classic. Be sure to look back at past styles and notice how the looks do not change.

9. Visit www.uspto.gov and conduct a search for "NASCAR." How many trademarks have been issued to this organization?

   The trademarks issued to NASCAR vary from teddy bears to beer mugs. Your list should be pretty long and comprehensive.

10. During your next visit to your local grocery store, locate a product that you think is a good example of brand extension. Justify your selection.

    Find a brand that offers several flavors of the same brand. Hunt's tomato sauce is a good example. Look at rice, pasta, peanut butter, and frozen dinners for more examples.

Chapter 8

1. Visit www.spriteremix.com and determine in which stage of the product life cycle this product is currently.

   Consider the use of technology on this site. This product is new, but is it still in the introduction phase?

2. Conduct a brief research study of the lodging industry. Explain why there are so many smaller hotels being acquired by the larger hotels.

   Use a site such as www.ahma.org to look at recent acquisitions and mergers. There are numerous other sites that track trends in the industry.

3. Compile a list of 5 products that only lasted months on the open market. How would you explain their demise?

   Think about products that you have purchased and then went back to the store to find with no avail. Consider what made that product special for you but may not have been to others.

4. Brainstorm ideas for new products with 2 friends and then again with 2 older adults. Is there a significant difference in concepts? Explain.

   There should be a difference in the types of products the older adults see as being viable than the younger perspective. The older adults will probably look at products that make simple tasks simpler, while the younger people will focus on technology that has not been tapped into.

5. Interview the manager of a local fast food restaurant and ask how the company develops ideas for new products.

   Inquire about the company's research and development policies. The manager may not be inclined to give out company secrets, but they should offer some general information about the process.

6. Identify 5 examples of print ads where the image and concepts seem mismatched to the product. Justify your selection.

   Consider any advertisement that shows images of anything BUT the product for these examples.

7. Go to www.lasvegas.com and find 5 examples of hotels/casinos that use virtual reality to promote their properties.

   Click on almost any hotel/casino to find a virtual tour.

8. Using your favorite search engine, find 5 websites that show prototypes for new products.

   Automobiles, palm pilots, and computers are excellent examples of products where new versions are heavily promoted.

9.   Using your favorite magazines and websites, find examples of products that differentiate between a style, a fashion, and a fad.

        Items that are considered in style are those items that you see in more than one place. A fashion trend will be featured in one magazine but not another. Clothing is a good start on this question.

10.   Find 3 websites that represent a company in the maturity phase. What are they doing to promote growth instead of decline?

        Think about companies that have been around for years (i.e. Coca-Cola, Microsoft, GE, and Ford) and look at the newer concepts and ideas they are generating.

Chapter 9

1.   Visit www.sheraton.com and search for a room in Miami, Florida during the month of June. How do you account for the variety of room rates?

        Consider the variety of choices you have when looking at a room. The rate should vary by the number of beds, the amenities in the room, and the view from the room. Also consider the discounts offered to various groups.

2.   Identify 5 products that you think are priced too low. Rationalize your answers.

        Answers here may vary based on your perception of what is "too low." You will want to justify your answer by associating a value to this product.

3.   Identify 5 products you think are priced too high. Rationalize your answer.

        The answers here will vary based on what you think is "too high." Sports cars are a good example. $65,000 may not seem like a great amount of money for something that is rare and hard to find.

4.   How do non-profit organizations price their products? If profit is not the basis, then what is?

        Most non-profit organizations look to cover their costs and some extras (like special events.)

5.   What enticed you to buy your last impulse item? Outline the justification for your purchase.

        Most of us impulse buy something we think we "have to have." We either fall in love with it OR it is so inexpensive we cannot pass up such a great deal!

6.   Conduct a brief research study on Mexican labor laws. Why are so many of our manufacturing companies relocating to Mexico?

        Look at www.mexicanlaws.com or www.nafta.com for more information on this topic.

7. List 5 products that are priced high at this moment in time, but the price will drop over the next 5 years.

    Consider the latest technology. Flat screen plasma televisions, Palm Pilots, etc. would be good examples.

8. Create a poster with 5 sections with pictures to represent each of the following: pure competition, monopolistic competition, oligopolistic competition, and pure monopoly.

    Your examples are your interpretation of what the meanings of these terms are. Use your textbook to help you pinpoint a good working definition of each term.

9. Create a list of items where you think demand is elastic and another list where you think demand is inelastic. Draw a comparison between the two lists.

    Consider those items where people are willing to pay anything for the product (cars, cosmetics, and clothing) and items where people will only pay so much (toothpaste, paper towels, garbage bags.)

10. Go to www.eddiebauer.com and conduct a search of their sale items. Outline how this strategy is an example of high-low pricing.

    Eddie Bauer has a section on their website that showcases their "clearance items." Notice how they are priced and the order in which they are showcased.

Chapter 10

1. Interview the manager of your favorite fast food restaurant. Outline the manner in which they receive products and ask them to determine the pros and cons of the method.

    When speaking with the manager, take into consideration the volume of product they move each day. Do they have storage for the inventory or do they get numerous smaller shipments instead of 1 big one per week?

2. You are trying to decide whether or not to custom order a car from overseas. What factors would you consider?

    Your first consideration is most likely cost. How much would each feature cost? How much is shipping? Taxes? Etc.

3. Compare the price of an item online at www.qvc.com to the regular retail price. How can you explain the difference?

    Consider the point in which the price of something goes down. Would the cost of the item go down if there were many items available?

4. When would there be a conflict between the prices of items in a catalog and the price of the item in the store? Explain.

   Consider the availability of the product and the ease with which the item is delivered to the customer.

5. How do rental car agencies deliver their cars to the customer? Conduct a brief research study and outline their system.

   Check sites such as www.hertzrentalcar.com and www.budgetrentalcar.com for information about the delivery of fleets.

6. Visit www.dol.gov and determine which job markets are deemed the most secure in the nation. Give a brief justification for the top three job markets in your area.

   There is a searchable database on this site. Select a specific area about which to find information. You do not want to get burdened with too much information.

7. Visit www.verabradley.com and discuss why this company is a good example of an exclusive distribution network.

   From this site, you should be able to locate a provider of Vera Bradley products in your area. Contact this site and ask the salesperson how they were selected to sell these products.

8. Outline the U.S. Trade Commission's procedure an individual would have to follow if they wanted to trade with the U.S.

   Go to www.usitc.com and locate the procedure. This site offers a variety of sources of U.S. trade law.

9. Investigate and identify 5 companies that offer international carrier services. How many ways can most companies transport goods?

   Use Google.com to locate international carrier services. www.maersksealand.com is an excellent example.

10. How do restaurants keep their inventory at a functional level? Interview a restaurateur in your area and find out.

    Consider the size of the restaurant and the market they service. The size of the inventory needed to maintain a par level would be one outstanding factor in determining how much to keep on hand.

Chapter 11

1. Do you prefer to shop in a retail store or in a non-retail store? Justify your answer.

   Your answer here will differ from that of the student next to you. Some people like the variety of a retail store and the heightened level of customer service they receive. Some shoppers prefer just looking without any assistance.

2. Using the three levels of service (self-service, limited-service, and full-service), find 3 examples of each.

Consider all the times in the past week in which you received service. The degree to which you had to help yourself is key to differentiating between the levels of service.

3. Visit a local store that has recently renovated its site. Will the changes have a positive effect on the store? Why or why not?

Many fast food restaurants, convenience stores, and grocery stores are getting a facelift. Notice any new services, signage, or conveniences they have offered.

4. Visit www.dollargeneral.com and give several examples of this store as a discount store.

The Dollar General is a great source of good bargains. Focus on those items you would normally buy in the grocery store and compare availability and prices.

5. Research Target and determine why the company decided to "super size" its stores.

Use sites such as www.target.com and www.fortune500.com to find information about company strategy.

6. Using your favorite search engine, find out all you can about Stanley Tanger. What was his vision and has it come to fruition?

Use a search engine to find information about Mr. Tanger. You might also try www.tangeroutlet.com for information about his company.

7. Novelty items such as sports fan paraphernalia are a common draw for a small target market. List the pros and cons of adding these items to your inventory.

Your answer here will depend on where you are and how large a role sports plays in your community. In South Carolina, college football is a beloved sport as is Nascar. How about in your neighborhood?

8. Visit www.ToysRUS.com and describe the atmosphere the website tries to create. Can atmosphere be created successfully?

When selling toys, what image do you want to project? How much fun do you want the buyer to have at your sight?

9. Using your favorite search engine, locate a large mall in your area. How many stores are included in this mall? How many different activities besides shopping are featured?

Consider all the different markets the mall is trying to appeal to. What are the people doing who are not shopping while they are waiting on the people who do?

10. Go to www.morganstanley.com and explain how this company could be considered a wholesaler. What is the firm selling and how?

Consider how this broker is making money. How do you turn a profit by selling something at the price it is worth?

Chapter 12

1. Find 10 examples of websites that use color and animation to attract children to their products.

   Look for websites that include cartoons and sound. You can look in the obvious places such as toy stores and theme parks, for examples.

2. Visit www.bandaid.com and show how Johnson & Johnson sells Band Aids to a large variety of customers.

   Band-Aids come in a variety of shapes, sizes, colors, prints, and textures. Who are all these products designed to attract?

3. Outline the 9 elements of the communication process in 5 30-second commercials. How do the different companies relay their messages differently?

   There will be similarities among all the commercials in the use of the communication process. The big difference will be in the way the message is encoded.

4. Give 5 examples of products that are more effectively marketed in a print ad than on the radio.

   Consider products that you need to see to get a feel for. Some products can be "pictured" with a verbal description and some items need to be seen to be believed!

5. Pretend you are going to buy a new flat screen plasma television set. Outline your internal dialogue through each of the 6 stages of buyer readiness.

   What questions would you ask yourself about this product? How would you talk yourself into this purchase? The first question most people ask is "How much is it?" or "Can I afford it?"

6. Locate 5 websites that use the "slice of life" style in their advertising. Justify your selections.

   Consider sites that show people in their everyday lives, doing everyday things. Also consider ads that show people living a lifestyle other than yours.

7. Why is the Super Bowl such an attractive time slot to premier new commercials? What is the average cost of a 30-second time slot during halftime and is this cost justified?

   Locate some statistics that reflect the viewership of the Super Bowl. Nielsen ratings would also be helpful here.

8. Identify a company that uses a celebrity to sell its product. Do a brief biographical sketch of this person.

   Celebrities are selling everything from cell phones to feminine hygiene products. But, there are also non-celebrity types that are just as effective. Consider Jared at www.subway.com.

9. Visit your local cosmetics counter and inquire about their "gift with purchase" sales incentives. Does the salesperson see a marked jump in sales during that time?

Clinique, Lancôme, Elizabeth Arden, and Estee Lauder all offer a "free gift with purchase incentive at different points during the year. You may also want to consider fragrance incentives.

10. Recall a human interest story that prompted you to buy something or make a contribution. Why was this story different from the countless others we are exposed to?

There are tons of products on the market now that are built around a cause. Breast cancer is an excellent example. The pink ribbon campaign offers a myriad of products where the proceeds benefit breast cancer research and awareness.

Chapter 13

1. Identify five alternative names for a salesperson.

Think of all the people that sell a product. Customer relations person, associate, employee, customer service representative, etc.

2. Visit www.monster.com and view the list of sales job openings in your area. Identify a company that shares your philosophy on personal selling.

When selecting a company, be sure to consider your personal selling style and your interests. Look at companies that fit your personality.

3. Contact the circulation desk at your local newspaper and inquire about how they find new customers. Explain how they divide their target markets.

Every newspaper office is challenged with selling new subscriptions. Does the company look at demographic and/or geographic information? If so, how is it utilized?

4. You are the manager of a phone bank charged with selling new long distance phone services. Script out a sales dialogue for your employees to follow.

How will you get the customer to stay on the phone? That seems to be the most common problem with telemarketers. Strategize how you make your product more attractive than the countless others.

5. Your boss offers you a job selling furnaces for straight commission or dishwashers for a salary plus commission. Which would you choose and why?

The answer here depends on your ability to sell and the frequency in which customers would buy your product. Your answer may vary from student to student, depending on the geographic area you live in and whether or not you feel comfortable selling one product over another.

6. Besides cash bonuses, what other kinds of incentives do you think would motivate employees?

Consider what makes you want to do something. Maybe gift certificates, day care coupons, paid time off, use of a company car or condo, etc.

7. Ask the salesperson at your favorite retail store how his/her performance is evaluated. Do sales goals play a role in determining whether or not he/she is successful?

Most salespeople have daily, weekly, or monthly goals they work toward. Whether or not meeting those goals plays a role in a performance evaluation may differ from company to company.

8. You are selling encyclopedias door to door. Outline your sales technique using all the steps in the selling process.

Refer to the steps in the selling process in your text.

9. Identify 5 companies that use direct marketing as an effective sales tool. What makes their use of direct sales more successful than others?

Consider the tool (brochure, hat, cups, matchbooks, etc.) and determine why this tool is more effective than another company's use of a similar tool.

10. Videotape your favorite infomercial and be prepared to share why you think it is a great selling tool.

There are hundreds of great infomercials out there. Pick one that really appeals to you as a buyer.

Chapter 14

1. Visit the web site for the United States Postal Service and discuss the significance of offering online services to customers.

Consider the variety of services being offered online, especially to business owners. Make a list of those services and then determine whether or not these would hold any "value."

2. Visit www.qvc.com and compare the use of the "marketspace" with the use of the television "marketplace."

Shopping online is very different from shopping on television. Generate a list of pros and cons of each method.

3. Identify 5 examples of online companies that have enjoyed a quick rise to success. Outline why you think these companies have been successful where others have not.

Use your favorite search engine to find information about "dot-coms." Most companies that have found quick success generated a good volume of news.

4.   Calculate the cost of running an interactive website for a year. Is this a cost effective method of doing business? Why or why not?

The answer here will vary from student to student. Whether or not the website would be cost effective would be contingent upon what is listed as a cost.

5.   Locate 5 restaurants that use the Internet to bolster their business. What techniques do they use that make their sites unique?

Most mainstream chain restaurants have websites. Try www.joescrabshack.com for a fun example.

6.   Determine whether eBay would be considered to have a "Business to Business" marketing strategy or a "Consumer to Consumer" strategy. Justify your answer.

Consider whether or not you consider eBay to be the consumer or the business and whether or not the seller is considered to be a business or a consumer.

7.   Locate 5 websites that fit each of the following criteria: college football, women's apparel, beer, news, and golf.

A search engine such as Google will be helpful in finding websites matching each of the above criteria.

8.   Visit www.gap.com and discuss the benefits of shopping online versus visiting the actual site.

Each student will have a different answer depending on how they like to shop. Some may like the physical experience of touching the merchandise and trying it on and some may prefer the convenience of home.

9.   Make a list of the "pop-up" ads you encounter in an hour of Internet surfing. Are these ads a selling tool or a nuisance?

Consider if you actually read the ads and if you do, why? What about a certain pop-up makes you stop what you are doing and read?

10.   Visit www.nyse.com and research the stock value of AOL Time Warner. How do you account for the variances over the past few years?

Take into consideration the mergers and acquisitions that have taken place over the past 5 years.

Chapter 15

1. Locate a governmental study on how foreign imports affect U.S. jobs. Pinpoint specific statistics for your state.

   Use governmental agencies, such as the World Trade Organization, for assistance.

2. Go to www.wto.org and locate a list of countries that have signed GATT so far. Who is missing?

   Use your knowledge of geography and your powers of deduction to answer this one.

3. What is the primary difference between the European Union and GATT? Find and print a list of the fundamental rights of the members of the EU, using your favorite search engine.

   Consider using Google to find information about the European Union. On the Union's homepage, there will be a site map to direct you to the pertinent information.

4. List the primary factors involved in determining duty rates.

   Use www.customs.gov as a resource.

5. Make a list of the natural resources of Iran. What is its primary export? Import? How could this information be a useful marketing tool?

   There are numerous websites dedicated to geographical information, but one of the most helpful resources is a good old-fashioned encyclopedia.

6. Locate an online currency converter and calculate the value of $1,000 in France, Germany, Spain, Bolivia, China, and South Africa.

   There are numerous websites that allow currency to be converted. Try www.xe.com for starters.

7. Go to www.export.gov and determine the number 1 export for the United States.

   Be sure to look at the documents posted on this site that report export information from all over the world.

8. You have been asked to create a baby doll for sale to children in the Middle East. What features would be different from that of an American doll?

   Consider the cultural differences between how women are viewed in the U.S. versus the view of women from a Middle Eastern perspective. Look at things such as clothing and accessories.

9. Identify 5 examples of current products that would have to change something about the packaging in order to be sold in a foreign market. Justify your selection by listing the product, the country in which it is meant to be sold, and the cultural issue that needs to be addressed.

Consider finding a website or a book in your school's library that deals with cultural issues. Once you know what you are looking for it should be easier to find. Student answers here will differ based on the products they select.

10.   Go to www.ebay.com and search for products sold by international sellers. What types of products are prevalent?

eBay has a huge section of items that are sold only by international sellers. Look for items such as books, videos, and clothing.

Chapter 16

1.   Make an argument for or against an additional sales tax on alcohol. Do you see this as a punishment or a way to generate revenue? Explain.

Use your personal opinion to form this argument. The basis of an ethical argument is what the person feels is right or wrong.

2.   Does your college bookstore have a captive audience in you? Is it an ethical issue for you that the bookstore can determine the price of books without justification? Explain.

Consider whether the convenience of the bookstore is a perceived value to you or do you see the bookstore as the enemy that holds hostage one of the essential tools you need to be successful in life?

3.   Go to www.congress.gov and conduct a background search on the Wheeler-Lea Act. Outline the circumstances that preceded this bill.

There is a search feature on this site that will allow you to find the information you need.

4.   Debate the ethical issue of a "pyramid scheme." Should these be illegal? Why or why not?

Consider finding a good working definition of exactly what a pyramid scheme does and why they are so controversial. You may also want to look and see if your state has a law banning these schemes.

5.   How did Ralph Nader become a fierce consumer advocate? Give a brief biographical background.

Use your favorite search engine to find out how Ralph Nader rose to such power and fame.

6.   Formulate a list of 5 organized consumer movements and outline each group's primary mission.

Consider looking at organizations that use a cause to prevent the purchase of products. PETA (People for the Ethical Treatment of Animals) would be a good example.

7.   List 5 websites that you consider to be examples of "green marketing."

Answers here will vary according to what each student finds.

8. Go to www.enron.com and find a copy of its "Code of Ethics." Give 5 examples of how the fall of Enron can be contributed to specific breaks in its code.

    There are numerous news stories that outline the ethical breaches made by the leaders at Enron. You may have to search an article database for your outline or you can draw conclusions based on your own interpretation.

9. Go to the website for your state legislature. How many bills are currently being considered under the title of "ethics"?

    Most states have a website that outlines what is happening on the legislative front. For example, South Carolina has an extensive information database at www.scstatehouse.net.

10. You are the CEO of a baby food company and have discovered that 100 cases of baby food may contain a mild form of salmonella. You shipped 1,000,000 cases of product last week alone. What do you do?

    Answers here will vary depending on the student. Consider the percentage of cases that may or may not be contaminated. Is a small outbreak worth recalling such a large amount? Formulate your list of potential problems and then weigh out the value of recalling the product.